D0559655

SEXUALITY AND MEDICINE

Volume II

PHILOSOPHY AND MEDICINE

Editors:

H. TRISTRAM ENGELHARDT, JR.

Center for Ethics, Medicine, and Public Issues,
Baylor College of Medicine, Houston, Texas, U.S.A.

STUART F. SPICKER

School of Medicine, University of Connecticut Health Center,
Farmington, Connecticut, U.S.A.

VOLUME 23

SEXUALITY AND MEDICINE

Volume II: Ethical Viewpoints in Transition

Edited by

EARL E. SHELP

Institute of Religion, and
Center for Ethics, Medicine, and Public Issues,
Baylor College of Medicine,
Houston, Texas, U.S.A.

D. REIDEL PUBLISHING COMPANY

A MEMBER OF THE KLUWER  ACADEMIC PUBLISHERS GROUP

DORDRECHT / BOSTON / LANCASTER / TOKYO

Library of Congress Cataloging-in-Publication Data
(Revised for vol. 2)

Sexuality and medicine.

(Philosophy and medicine ; v. 22–23)
Includes bibliographies and index.
Contents: v. 1. Conceptual roots – v. 2. Ethical viewpoints in transition.
1. Sex (Psychology)–Collected works. 2. Medicine and psychology–Collected works. I. Shelp, Earl, E., 1947- . II. Series.
BF692.S4347 1987 306.7 86-26201
ISBN 90-277-2290-0 (v. 1)
ISBN 1-55608-013-1 (v. 2)

Published by D. Reidel Publishing Company,
P.O. Box 17, 3300 AA Dordrecht, Holland.

Sold and distributed in the U.S.A. and Canada
by Kluwer Academic Publishers,
101 Philip Drive, Norwell, MA 02061, U.S.A.

In all other countries, sold and distributed
by Kluwer Academic Publishers Group,
P.O. Box 322, 3300 AH Dordrecht, Holland.

Printed in The Netherlands.

TABLE OF CONTENTS

Volume II

TABLE OF CONTENTS
Volume I

FOREWORD

It may be unnecessary to some to publish a text on sexuality in 1986 since the popular press speaks of the sexual revolution as if it were over and was possibly a mistake. Some people characterize society as too sexually obsessed, and there is an undercurrent of desire for a return to a supposedly simpler and happier time when sex was not openly discussed, displayed, taught or even, presumedly, contemplated.

Indeed, we are experiencing something of a backlash against open sexuality and sexual liberation. For example, during the '60s and '70s tolerance of homosexual persons and homosexuality increased. Of late there has been a conservative backlash against gay-rights laws. Sexual intercourse before marriage, which had been considered healthy and good, has been, of late, characterized as promiscuous. In fact, numerous articles have appeared about the growing popularity of sexual abstinence. There is a renewed vigor in the fight against sex education in the schools, and an 'anti-pornography' battle being waged by those on the right and those on the left who organize under the guise of such worthy goals as deterring child abuse and rape, but who are basically uncomfortable with diverse expressions of sexuality.

One would hope that such trends, and the ignorance about sex and sexuality that they reflect, would not touch medical professionals. That Dr. Earl Shelp has spent years gathering these essays on sexuality and medicine only goes to underscore the sad reality that medicine is not immune to these regressive trends. Professor Shelp's interest in this collection grew out of observations during clinical work that many concepts of human sexuality were poorly understood by faculty and, consequently, students were either kept in ignorance or misinformed. Moreover, he found that historical and conceptual judgments concerning human sexuality tended to be myopic as clinicians perpetuated the received tradition uncritically. Consequently, what are taught as medical truths may actually be religious, philosophical, or cultural precepts, not based on scientific reality.

Earl E. Shelp (ed.), Sexuality and Medicine, Vol. II, ix-xi.

This comprehensive text calls to the attention of physicians the relationship of conceptual understandings to normative judgments in the area of sexuality. In spite of the sexual revolution, it appears that in the main today's medical students and practicing physicians perpetuate arbitrary judgments about normal and abnormal sexuality; they are highly judgmental about practices that do not conform to their own; and they are ignorant of the variety of possible human sexual expression. The devastating results of these attitudes are obvious throughout our culture. Moral judgments about sex should not be expressed as scientific absolutes in a pluralistic society that values self determination and individual freedom.

This fundamental American belief in individual freedom is one that has concerned my father, Hugh Hefner, for thirty years. It is expressed in *Playboy* magazine and in the work of the Playboy Foundation. In the early 1960s, a series of editorials penned by him and termed the 'Playboy Philosophy' offered a historical overview of individual freedom, particularly sexual freedom, and argued passionately for the right of individuals to be free of governmental or religiously imposed morality.

Through the Playboy Foundation, Hefner put his money where his mouth was. It made the initial grant to establish an Office of Research Services of the Sex Information and Education Council of the U.S. (SEICUS) in the late '60s. And in 1971, Playboy awarded a grant to establish a pilot program at the University of Minnesota to develop curricula for changing the attitudes of men and women medical students. The premise of the program, as stated in the grant proposal, was:

> While the physician is most frequently consulted in sexual matters, he is woefully ill-prepared to handle them. The recent graduate of medical school is often in the difficult position of an 'expert' who knows *less* about his subject than his patient. Medical students, by nature of their social and economic background, personality types, and scientific training, are traditionally among the most inhibited, threatened, naive and anxiety-ridden of professional students. Unfortunately, even the physician who is factually knowledgeable about human sexuality is often unable to use the information he possesses due to his emotional response to the subject. Our premise is that much of any attitude toward sexuality is based on myths regarding sexual behavior, and that perpetration of these myths leads, at best, to unsound therapy and counseling, and at worst, to failure of the M.D. to even acknowledge sexual problems.

The rationale for these volumes suggests that the state of medical practice today is not much better than it was in 1971. Perhaps this is due to a still present uncomfortable titillation that permeates medical interest in sexuality. In the '60s, when the Playboy Foundation became the major research sponsor of the Masters & Johnson Institute, we did so, not only because of the importance of the work, but also because it was virtually impossible for the Institute to attract more traditional support due to this attitude toward sexuality. Tragically, the Institute's situation remains substantially similar today.

This superstition and embarrassment, which continues to cloud our understanding of human sexuality, makes it critically important that medical professionals, as well as social scientists, philosophers, theologians, and the general public be exposed to the insights and information on human sexuality that Professor Shelp has developed in this book.

I hope that this text is widely used, since its subject is so important to each of us.

Playboy Enterprises, Inc.,
Chicago, Illinois, U.S.A.

CHRISTIE HEFNER

PREFACE

The statement that societal sexual norms are in transition implies popular disagreement about the propriety and morality of sexual behavior. It also may signal confusion in the law. Present disagreements about sexual behavior reflect confusion over what is right for individuals and society and, moreover, what is good, i.e., contributes to the health of individuals and of society. One question that is rarely explored in such times of flux, and one that this volume attempts to address, is, what sort of authority is appropriate to resolve the inevitable tensions of change?

For example, various professions have proclaimed the relevance of their expertise regarding the proper attitude for society to adopt in a time of moral transition. Sociologists, political scientists, lawyers, and theologians have all identified possibly negative societal consequences of individual behavior. These same pundits argue for appropriate social responses, which may mitigate their predictions. Other professionals, therapists, sex counselors, physicians (and some cult figures) have intervened in the discussion in other ways; they propose to counsel, comfort, or cure – to integrate sexuality into personality in a way that maximizes harmony and minimizes discord. Congress has demonstrated its disapproval of teenage sexuality by attempting to limit access of these persons to contraceptive information and devices, thus raising a voice in favor of abstinence – a somewhat bizarre response to the increasing incidence and prevalence of teenage pregnancy. At the same time medical clinics proliferate whose object it is to provide prospective advice and devices for sexually active teenagers and for active homosexuals. Each group is convinced of the benefit to individuals and to society from its approach.

In all of this confusion, one could ask how we as a society can best grapple with defining and protecting society's legitimate interests, while permitting individual expression. Can society establish and enforce norms that constrain an individual's choice of person or method of

Earl E. Shelp (ed.), Sexuality and Medicine, Vol. II, xiii-xx.

satisfying her own needs and desires? Answering this question is particularly perplexing in matters relating to sex: most individual sexual behavior is private and hidden. Public debate is concerned with many aspects of the morality of sexual behavior: first, for consenting adults; second, the debasement of behavior in rape and forced actions; third, issues of child involvement or abuse. In all of these, the unitary question can be asked: what are the consequences of freely assumed and forced sex for society? The next question could be, which professional intervention promotes the greatest good for an individual and for the community?

Despite current fondness for decrying the medical model, it does provide some measurable benefits for the individual and, I would argue, for society in this time of debate and uncertainty. One benefit of applying a medical model to issues of sexuality is that it provides for an exclusive focus on the individual and her needs. The competing and conflicting needs of other individuals and society are secondary, if recognized at all. Care for the body is historically the province of medicine. In the medical arena, socialization provides that less embarrassment is likely to encumber the discussion. In addition, medical caregivers can offer concrete interventions with no legal risk to themselves or to their patients; they act to control the spread of sexually transmitted disease; and they can promote information about reproduction and enhance the autonomous decision of a woman as to whether to bear or beget a child.

The usual argument against the medical model, that it stamps the discussion with inappropriate labels of sickness and health, has lessened validity in this discussion. Those valid critiques relate more to the problems of professional aegis than to individual caregiver perspectives. In a perverse way, the private patient-centered, confidential practice of medicine provides a shield for individuals whose behavior is deemed by some to be immoral or illegal. Medicine, rather than labeling an action as unhealthy, may thus be the best positioned to respect individual rights.

One could argue that this medical protection is inappropriate. As sexuality is central to human functioning and to the propagation of society, individual actions are of proper concern to society and to the professional subdivisions of society charged with defining, enacting, and implementing public policy with regard to sexual behavior.

This argument underlies some of the contests presently emerging between medicine, lawyers, and legislators regarding which set of assumptions and skills is best suited to address issues of individual sexuality in the context of community. Is sexuality of exclusive concern to an individual and that person's chosen caregivers? Is it a matter of care for a body? Is what is at issue the sort of contraceptive device, or the techniques for controlling the spread of herpes or other venereal diseases? Are sexuality and individual sexual conduct a matter of singular concern for the provider and patient? Can we distinguish between sexual function and sexual behavior? Is that a useful distinction to divide private concern and public position?

Where should the proper locus lie for the consideration of these and related issues? A woman's right to reproduce, that is, to conceive and abort, may be of more interest to society than her right to couple and cavort. The two are, however, so inescapably intertwined that the regulation of the first may argue for an interest in limiting the second. A teenager's right to sexual intimacy challenges parental authority. Patterns of teenage sexuality may reflect profound changes in family, economic, and educational structures in society. It may also mirror evolving norms of behavior and shifts in assumptions about the role of children. It occurs, however, amid reports on the increasing incidence and prevalence of child abuse. Is there some connection? Data on increased incidence of the sexual abuse of children may result from greater awareness or may represent a real increase in the problem. Are such data more relevant to family dysfunction or to norms of sexual behavior?

Societally sanctioned norms of behavior may be at odds with the working knowledge about needs, wants, and desires of certain individuals. This is especially true in regard to health provider practice and advice in matters of sexuality. This tension between the assumptions and imperatives that shape medical practice and the vision and constraint imposed by society is occasionally revealed and reflected in legal opinions. A doctor-patient relationship should, by the ethic of practice, be fueled only by considerations of patient need. Appropriate public policy constraints on this relationship are few. Mandatory reporting of violent crime, of infectious disease, or of child abuse represents the most common example of actions taken by the physician against what the patient may define as self-interest. The doctor-patient relationship is individually defined in specifically crafted contracts of care. In contrast, laws, legislators, lawyers, and judges address individual action in the frame of society. Sexuality is an individual matter. Nonetheless, patterns of sexuality paint colors in the fabric of society.

Case law opinions are, as Justice Holmes commented, both a response to the unconscious preferences of society and a reflection of public policy. Decided cases are also, however, representations of the personal preferences of individual jurists within the confines of larger political debate. As an illustration of these principles in this discussion of professional aegis and sexuality, and the role of law and medicine with regard to defining acceptable conduct, consider *Dronenberg v. Zech* [1], a case presented to the United States Court of Appeals for the District of Columbia and decided in an opinion by Justice Robert H. Bork.

The facts of the case were not in dispute. James L. Dronenberg was discharged from the United States Navy for repeated homosexual conduct. The discharge was permitted by regulations stating in part that:

> any member of the Navy who solicits, attempts or engages in homosexual acts shall normally be separated from the naval service. The presence of such member in a military environment seriously impairs combat readiness, efficiency, security and morale [1].

Dronenberg acknowledged engaging in homosexual acts in the Navy barracks. The Navy Administrative Discharge Board voted discharge and the Secretary of the Navy, on review, ordered the discharge to be characterized as honorable.

On appeal, Dronenberg's case advanced two constitutional arguments, a right to privacy and a right to equal protection. The second of these was, however, dependent on resolution of the first. That is, if no right to privacy protecting homosexual behavior exists outside the military, then doctrines of equal protection would not compel extension to persons in the armed services.

The primary question considered by the court was whether there exists a constitutional right to privacy based on fundamental human rights, substantive due process, the Ninth Amendment, or emanations from the Bill of Rights. Second, the court asked, if this right exists, does it protect the sexual activity of consenting homosexual adults not engaged in military service? If it does, the court stated, the protection could then be required to be equally available to military personnel.

The appellant Dronenberg argued that previously decided Supreme Court cases presented a thread of principle, to wit, that "the government should not interfere with an individual's freedom to control intimate personal decisions regarding his or her own body except by the least restrictive means available and in the presence of a compelling state interest" [1]. Private consensual homosexual activity, he argued, is within the zone of constitutionally protected privacy.

The opinion of the court reviewed the major cases relating to the right to an abortion and the right to use contraceptives. Both of these are activities that were specifically protected by this right to privacy in previously decided Supreme Court opinions. It then proposed its view that there is no independent fundamental right of access to contraceptives or right to use contraceptives, but rather constitutional barriers to interference with access because of the constitutionally "protected right of decisions in matters of childbearing that is the underlying foundation of the holdings" (in the contraception and abortion cases). The court concluded that "we would find it impossible to conclude that a right to homosexual conduct is 'fundamental' or 'implicit in the concept of ordered liberty' unless any and all private sexual behavior falls within those categories, a conclusion we are unwilling to draw" [1].

This absolute refusal to discover an applicable principle could provide a basis for substantive analysis or a foundation for a philosophical or jurisprudential critique. Ronald Dworkin, writing in the *New York Review of Books*, concluded his analysis of the case as follows:

> This is the jurisprudence of fiat, not argument. It is sadly consistent with the Republican platform, which calls for the appointment of judges 'at all levels of the judiciary' with the proper, that is to say right-wing, views about 'traditional family values and the sanctity of innocent human life,' as if only a justice's personal moral convictions, not his or her arguments of law, really matter. If justices with that view of their work colonize the Supreme Court, earning their places through decisions like Bork's, the Court will no longer be what our traditions celebrate, forum of principle where unpopular minorities can argue for liberty on grounds of right. It will become the Moral Majority's clubhouse, where the prejudices of the day are called constitutional law [2].

If Dworkin's analysis is correct, and if the judiciary comes increasingly to reflect personal morality and 'right wing' politics, medical caregivers committed to individual patients are likely to be in a pattern of practice at odds with the assumptions of public policy makers and politicians. They are also likely to be in the best position to protect individual interests even if these are not protected by law. This will be so because traditions of confidentiality and fidelity to patients will protect individual communications and the practices they reveal.

This protection by medicine must be weighed against the dangers of imposing a medical model on complex social problems. The medicalization of aging and long-term care is the example perhaps most prominent at present. By designating aging as a medical matter we have removed many elderly from society and placed them in institutions designed to treat and confine rather than enhance and support independent functioning. The association of medicine with the elderly has encouraged applying to them the label of "sick". Sick is bad; healthy is good. Old age is bad; youth is good. The net result of the hegemony of medical management may thus be to encourage disability and dependence in perception if not in fact. Another example of the misuse of medicine could be the role of psychiatry in the criminal justice process – an additional easy target for negative comment.

Consider, however, what may be some of the risks, benefits, dangers, and dilemmas of accepting a medical model for judgments of acceptability in the area of sexuality, and specifically in the area of homosexuality. Consider Mr. Dronenberg, for example. Does Mr. Dronenberg have a disease? Until 1979, at which time the American Psychiatric Association voted to remove homosexuality from its list of psychiatric disorders, Mr. Dronenberg was labeled "sick" in the standard medical model of practice. Due to the vagaries and politics of the evolution of nomenclature, he is now "well". Clearly all physicians do not respond equally to this shift in perspective. But certainly careful culling of medical practitioners will increasingly yield individuals sympathetic, supportive, and non-judgmental of homosexual persons and practices. Homosexuality as such, as approached by the organized profession, is no longer a medical problem. This is not to suggest that homosexual behavior is necessarily beyond the concern of physicians. Counseling about homosexual health and risks associated with homosexual behavior should be made available in the context of the provider-patient relationship.

Homosexuality as one variant of "private sexual behavior" is, given the Dronenberg case as an example, most definitely a problem for law and society. The reluctance of Judge Bork to recognize constitutional protections for sexual behavior of consenting adults may reflect a personal political philosophy and preference, as Dworkin argues. It may also reflect, however, some unarticulated sense of public policy – that the community has some legitimate stake in maintaining some aegis over patterns of private conduct. Dworkin argues that the developing Supreme Court concept of "right to privacy" required Judge Bork to enunciate and apply this right to Mr. Dronenberg's behavior. The legal-societal relationship to sexuality, as this case indicates, is not as individually limited and constrained as the medical model. Judge Bork could have argued that communal perception, through legislative action, is the only appropriate vehicle for extending societal sanction to all actions of consenting adults. The legal result may be the same. This argument, however, would change the plane of debate from personal preference to public principle.

It would have been far more helpful for Judge Bork to identify and provide arguments for an appropriate state interest in matters of sexuality and the right to privacy. Instead he appears to disregard the spirit, if not the letter, of the previously decided cases. As he did not approach the matter in this way, the case provides no guidance for defining the legitimate interests of community and society in these private behaviors.

Private sexuality may be protected not by concepts of right but by traditions of the doctor-patient relationship. The medical model is committed to supporting the personal preference and idiosyncrasy of individual patients. For example, doctrines of informed consent and confidentiality require communication with an individual patient and protection of that information (absent specific danger to society or an identified other) from society. These are patient-specific directives. These sorts of principles, rather than concepts of illness and health, may be the most important and helpful aspect of the medical model for the protection of individual sexual behavior. If health depends on sexual equilibrium, a medical model can comfortably protect behavior labeled "variant" by society.

The legal and societal norm is much more problematic for individuals. Law can empower; it can also constrain. In matters of sexuality it will most likely attempt to do both. The development of legal norms will be related to issues of politics as well as to considered and carefully defined community interest. In this context medicine, rather than law, may emerge as the protector of individual liberty.

Law and medicine both deal with individual behaviors on which moral judgments rest. This volume attempts to explore the perspectives of the professions in regard to definitions of legitimate societal interests. As such it is an important contribution in the process of honing societal morality.

Montefiore Medical Center,
Bronx, New York, U.S.A.

NANCY N. DUBLER

BIBLIOGRAPHY

1. *Dronenberg v. Zech*, 741 F. 2d 1388, D.C. Cir., 1984.
2. Dworkin, R.: 1984, 'Reagan's Justice', *The New York Review of Books* **27** (Nov. 8), 27–31.

INTRODUCTION

Few subjects seem to generate more intense and uncompromising debate than those focussed on human sexuality and sexual morality. The sources for normative judgments about these matters may be religious, secular, cultural, or some amalgam of these or other influences. A set of opinions, it appears, often is embraced by people without first critically reflecting upon their conceptual foundations or implications. As a result, pronouncements tend to be made about how medicine or physicians ought to respond to sexuality-related phenomena without clear understandings of medicine, physicians, or human sexuality. Important questions relevant to these judgments may never be asked or, if asked, answered inaccurately or not very satisfactorily. Thus, what one was taught or has come to believe about medicine, sexuality, sexual morality, and the role of physicians in society is presented as a valid, singularly correct statement based on what, in reality, may be insubstantial evidence. Examples of the sorts of pronouncements that require a justification that may not be given include: Physicians ought not perform abortions, masturbation is contrary to nature, homosexuality is a psychopathology, and transsexual people ought not to be treated surgically.

Some of the conceptual aspects of many of the troubling issues that connect the philosophy of medicine, theories of medicine, and bioethics are explored in the companion volume to this collection [1]. The analyses contained in that volume demonstrate the difficulty of attaining a consensus about the meaning of certain concepts (e.g., natural, sick, feminine, artificial, disease, healthy) relevant to discussions about human sexual function, human sexual expression, human sexual identity, and the proper relation of medicine to each. As importantly, the dispute about the meaning of basic concepts accounts, in part, for the vigorous disagreements about what is right or wrong regarding, for example, certain forms of sexual expression and the response of physicians to them. And finally, the relative weight of these conceptual

Earl E. Shelp (ed.), Sexuality and Medicine, Vol. II, xxi-xxxii.

understandings to normative judgments of sexual morality and medical practice are debated.

The essays in this volume focus on the moral debate about the relation of medicine and physicians to human sexual function, human sexual expression, and human desires, feelings, and complaints that are sexual in nature. The contributors' analyses reveal reproductive, relational, and recreational approaches to understandings of human sexuality and sexual morality. Moreover, they disclose how cultural, philosophical, medical, and religious presuppositions and viewpoints shape or influence moral and medical judgments of what is and is not a licit use of sexual organs, function, or expression. The foundation and shape of many of the arguments regarding the interrelation of morality, sexuality, and medicine are illustrated, analyzed, and evaluated in the process. The discussion of sexuality and medicine is presented in three sections. The first section contains four essays that address descriptively and critically views regarding the purposes of human sexuality, the role of medicine with respect to them, and the search for a secular sexual ethic that takes account of evolving understandings of human sexuality and medicine. Section two has six essays that focus on the influence of moral and non-moral values on medical and scientific judgments regarding human sexual function and expression. In addition, what these judgments are held to medically or morally prescribe or proscribe is critically examined. The third section of four essays explores the relation of religious beliefs and values to sexual ethics and medical practice. General religious views regarding sexuality and sexual morality are described and evaluated. Specific subjects of medical, moral, and religious interest also are discussed, e.g., transsexualism, homosexuality, masturbation, coitus outside of marriage.

The first section begins with an essay by Mary Ann Gardell. Gardell investigates the significance selected philosophers assign to sexuality, particularly as it relates to an individual's view of the good or moral life. She notes that understandings of sexuality and ethics are complex and conditioned by numerous variables. Nevertheless, four major traditions in Western philosophical thought regarding sexual ethics are identified and reviewed: (1) there is little of unique significance with respect to the nature of sexual ethics, (2) sexual ethics can be based in some concrete understanding of sexuality, (3) sexual ethics are derived from certain intrinsic features to morality, and (4) basic moral principles are to be applied to sexual issues. Her analysis of these traditions leads her to conclude that there is no unique sexual ethic. Rather, moral guidance is to be found in general principles like promise-keeping, truth-telling, and mutual respect. Accordingly, for Gardell, sexual ethics are "heterogeneous and complex . . ., fashioned by a wide range of claims, interests, desires, and goals that influence the ways in which we describe, explain, and evaluate sexual behavior."

Turning from a general survey of secular sexual ethics, Sara Ann Ketchum calls attention to and analyzes one complicating feature of the debate about sexual morality. She points out that reproductive decisions raise a class of interesting moral issues not raised by most other medical decisions. This is so primarily because an affirmative reproductive decision results in the creation of a new subject of moral consideration, the offspring, who nevertheless does not exist at the time the decision is made. How are the interests of this "prospective person" to be factored into the parents' reproductive decision? Ketchum discusses the many facets of this question, ranging from the moral legitimacy of genotype-altering cures for congenital defects to the moral responsibility of the present generation to produce a future one. The author argues that in most reproductive decisions the interests of the parents (and particularly of the mother) should override those of the prospective offspring. She concludes by observing that the medical profession's monopolistic control of reproductive technology has given physicians much more say in reproductive decisions than is justified by a consideration of their interests in such decisions.

Lisa Cahill's essay, "On the Connection of Sex to Reproduction," explores several interrelated subjects: human life, as individual and as communal; love, liberty, sexuality, corporality, reproduction; and the role of medical professionals as persons multiply interfaced with other persons seeking or needing medical advice about matters concerning, in particular, the control of reproduction. A vital and recurrent theme is best expressed in the author's own words:

> I think it not implausible to suggest that there ought to be a *prima facie* presumption in favor of the essential characteristics of human sex as normative for individual sex. If there is, though, this presumption certainly can be overridden by factors in individual situations which count against the successful integration of procreation into sexual relationships and commitments. Why? Because personhood is more distinctively human than biology (reproduction). In humans, freedom and rationality control and direct biology and the physical conditions of life in community. This is what is meant by saying that humans are moral agents at all.

Cahill is restive in her views, not dogmatic. Her essay is indicative of a deep and abiding interest, research, and reflection on sexuality-related issues, rather than an algorithm for solving concrete moral problems. Yet she offers, as her concluding section, a very suggestive display of her ethical methodology by considering a few of the startling moral questions posed by modern reproductive technology and the role of medical professionals in using it.

In the final essay of the first section, H. Tristram Engelhardt, Jr. observes that three clusters of goals are supportable by human sexuality (reproduction, bonding, and recreation) which can conflict. He proposes and defends the view that in such situations of conflict nothing morally unique is at issue in most cases, only interesting examples of how to apply common, general moral principles. Claims that norms for sexual behavior can be discerned from the design of nature are criticized. Instead, the author argues in favor of the view that goals for sexuality are provided by individuals or by particular moral and aesthetic traditions. Engelhardt proposes that "within friendships and families the desires and urges that characterize sexuality are provided interpretive matrices of ideas and values." As such, the fact of numerous sexual moralities ought to be expected and accepted. There are, however, general restraints on sexual expression, according to the author. These are the constraints of mutual respect which is the necessary condition for the possibility of resolving moral controversies on principle without force.

The second section on society, sexuality, and medicine opens with John Duffy's historical review of attitudes toward human sexuality and sexual ethics in America from the Colonial era to the present. This history reveals an evolution of public attitudes, beginning with a relaxed, tolerant, even positive view toward certain sexual practices in early America, and moving toward a rigid, intolerant, and fearful view by the middle of the nineteenth century. According to Duffy, much of this change may be attributable to the influence of the medical profession, which tended to assume a moral authority over sexual matters in lieu of its failure adequately to understand the physiology and medical pathology of sex. Beliefs prevalent in America from roughly the mid-nineteenth century until the end of World War I are represented by Duffy as highly repressive, particularly on matters of masturbation, female sexual physiology and behavior, prostitution, and venereal disease.

After World War I, Duffy notes a dramatic shift in many of these repressive attitudes. For example, sexual matters were again considered appropriate for public and professional discussion. Women moved more into the workplace and managed to alter at least a few people's opinion that the female gender is too delicate and too tender to be fully human. The idea of women as sexual beings with sexual needs of their own emerged at this time. An unfortunate side-effect of this liberalized attitude toward women was their exploitation by sexually-oriented film makers in the twenties. Following World War II, Duffy shows how American sexual attitudes became even more open. In particular, he discusses the important role of women in the American military effort against Germany and Japan, and their increasing status as status-rivals of men in American society. Duffy concludes with a brief analysis of the contributing causes and some of the effects of the latest sexual "revolution" in America which began during the decade of the 1960s.

Following Duffy's historical and cultural orientation to medicine's relation to sexuality in America, Robert Baker argues that physicians should be neither moralistic nor completely amoral when advising patients in sexual matters. Rather, physicians should be "sexual philosophers" creating an atmosphere of a "philosophical forum" in which patients feel free to explore with physicians their questions, beliefs, and concerns regarding sexual life-style. Baker gives two case studies, one involving masturbation and the other anal sex. He then presents four common responses and shows why each is inadequate. The first three, in his judgment, are too moralistic, and the fourth is too devoid of moral content. The first view is described as the "puritanical" view of Augustine. Augustine, according to Baker, asserts that the human will should control the body. Masturbation is wrong, accordingly, because in it the will becomes a servant of the body. The second view, elaborated by Thomas Aquinas, is that God ordained everything for a purpose. The purpose of semen, Aquinas believed, was procreation. Thus any other uses, such as masturbation and anal sex, must be wrong. The third view is Kant's view that persons ought to treat each other as ends and never as means. Only when there is "a unity of wills," Kant believed, could sexual intercourse meet these moral requirements. Naturalism maintains that morality should conform to nature and that natural urges should not be considered wrong. Baker cites examples of natural urges and actions that are indeed considered wrong, thus refuting the naturalist position. Baker urges a middle way between a moralistic, judgmental posture and an amoral one. Neither extreme meets the need of the patient, which is to discuss moral and sexual concerns with their physician in an open, direct manner.

The discussion then turns to a consideration of what sexual behaviors, if any, are psychopathological warranting medical intervention. Frederick Suppe argues that variant sexual behaviors are not indications of mental disorders, as is commonly believed. Their inclusion in the American Psychiatric Association's list of mental disorders, contained in the third edition of *Diagnostic and Statistical Manual*, is not warranted by empirical evidence. This suggests to Suppe that psychiatry may be less an empirical science than a pseudo-scientific "codification of social mores." To prove his thesis Suppe examines in detail two variant sexual behaviors, homosexuality and sadomasochism. He finds no indication of a connection between homosexuality or sadomasochism and mental disorder. He also dispels the myth that persons who engage in sadomasochistic sexual practices are more violent, criminal, or in any other way socially deviant than the rest of the population.

One of the main types of psychosexual disorders listed in DSM-III is the "paraphilias," which involve sexual arousal in response to objects other than the "normal" ones, and in which "the capacity for reciprocal affectionate sexual activity" may be impaired. Paraphilia include transvestism, zoophilia, pedophilia, sexual sadism, sexual masochism, urophilia (urine), and so on. Suppe challenges the claim that most persons who engage in these practices are truly paraphiliac. To be truly paraphiliac, he argues, these "abnormal" cues would have to be necessary for sexual arousal. For the vast majority, these cues are facilitative but not necessary. Suppe also challenges the claim that most persons who engage in paraphilia are incapable of reciprocal affectionate sexual activity. He believes that most can and do. The major barrier is not their inability to affectionately accept others but society's unwillingness to be affectionate and accepting towards them.

The next two essays turn away from matters of classification to examine how sexual values and life-styles are changing and the response of medical practitioners to these changes. Joshua Golden observes that our society has become less tolerant of child abuse and more tolerant of homosexuality, cohabitation, sadomasochism, children engaging in masturbation and coitus, and the use of such technological innovations as penile prostheses, artificial insemination, and surrogate pregnancies. Other trends include the general population's increased exposure to sexually oriented media, greater emphasis on sexual fulfillment through masturbation and forms of eroticism other than coitus, and an aging population that will insist on remaining sexually active. Golden examines the implications of these changes for medical practitioners. For example, he recommends that hospitals and nursing homes be designed to allow greater privacy for persons who, although sick or aging, will desire to remain sexually active and for whom being sexually active is essential to being healthy. Golden urges physicians to keep abreast of the changing values and life-styles of their patient population and to reflect upon the ethical implications of these changes.

This discussion of patient-physician interaction regarding sexual matters is extended by Nellie Grose and Earl Shelp who forward the view that sexual values and life-styles are significant factors in a person's overall health that too often are unattended by physicians. This failure to give appropriate consideration to the sexual aspects of human health can be traced to inadequate training at the residency level and to the tendency of physicians to assume a too authoritarian role in relating to patients. Grose and Shelp recommend that physicians take a stance as morally neutral as possible with regard to the sexual values and life-styles of patients, that they encourage their patients to discuss sexual problems with them, and that they learn to "read between the lines" of what the patient says . The authors argue that physicians should respect the sexual values and life-styles of their patients. When a physician disagrees with the patient's sexual values and life-style, the physician should not assume that the solution is to get the patient to alter his or her values and life-style. The physician should strive, as much as possible, to treat the patient within the patient's value system. They caution against the use of terms like "normal" and "abnormal," "moral" and "immoral," and "superior" and "inferior" with regard to sexual values and life-styles. Physicians should try to view their patients' sexual values and life-styles as aesthetic rather than moral choices.

Because physicians are persons with values and life-style choices as well, it is unrealistic to expect them to be completely neutral. Whenever values surface, they should be recognized for what they are, personal values and not "brute facts." Grose and Shelp illustrate this approach through six case studies: (1) a lesbian woman wishing to become pregnant by artificial insemination; (2) a widowed woman trying to resolve the conflict between her fundamentalist religious beliefs and her desires for masturbation and sexual fulfillment; (3) a man who wishes to remain sexually active but who has a history of sexually transmittable diseases; (4) a 50 year old woman whose husband has Alzheimer's disease and can no longer meet her sexual needs; (5) a woman who complains that her husband wants to experiment with group sex; and (6) a heterosexual male fashion model who believes that his career would be enhanced if he could perform homosexually.

The final essay in the second section reviews the history, methods, and moral problems of sex research and therapy. J. Robert Meyners claims that the importance of sexuality to human health makes sexual research and therapy necessary. But because of the intimacy of sexual experience, research and therapy becomes difficult, scientifically and morally. More specifically, ethical issues surrounding sex research and therapy touch deep emotional and religious sensibilities. Thus, Meyners concludes that guidelines are needed. He suggests that guidelines be based neither on one philosophical or religious system (too restrictive) nor on allowing each therapist to follow his or her own belief system (not restrictive enough), but on standards consistent with the relevant social and historical context. In the American context, these standards can be expressed in terms of the following principle: "no party shall be interfered with in their pursuit of their cherished values." This principle is applied by the author to the major ethical problems faced by sex researchers and therapists: the use of sexual surrogates, confidentiality between therapist and patient, and informed consent. He offers guidelines, for example, in determining which situations merit the use of surrogates, how to evaluate the competency of the surrogate, how to protect the welfare of the surrogate, how to respond to patients who ask therapists to keep certain infomation confidential from the patient's spouse or lover, and the role of informed consent in resolving conflicts over differing values of therapist and patient, and researcher and subject.

The final section of essays explores the relevance of religiously based viewpoints regarding human sexuality to sexual ethics and medical practice. The opening essay by Paul Simmons provides an overview of theological approaches to sexuality and explains how certain moral norms relate to basic theological affirmations. The author shows how theologians may reach similar conclusions about particular questions even though differing reasons are utilized, and, at times, how strong differences may develop on the basis of different beliefs. He examines basic assumptions of natural law, protestant fundamentalism, "orders of creation" theology, situations ethics, and process theology. In the process, Simmons reviews varying judgments regarding specific sexual acts and purposes of sexuality, providing a critique of the methodology underlying them. In a concluding section, Simmons notes that appeals to nature, scripture, rules, or philosophical perspectives, in similar or differing ways, can lead to diverse judgments about sexuality, each strongly advanced and defended by its proponent. This being so, Simmons calls for mutual respect and an openness to unfamiliar perspectives as a way to discern God's redemptive purposes for human sexual nature.

James McCartney next surveys the development of Roman Catholic moral theology since Vatican II, characterizing it as a dialectical interaction between two conflicting approaches that are both essentially teleological. He describes these approaches as "prudential personalism" and "proportionalism." He illustrates how both methods of moral analysis are applied to areas of controversy (masturbation, homosexuality, premarital sex), showing not only how each camp reasons, but the differing judgments which result. Proportionalists' judgments tend to include a consideration of the context and the act, whereas prudential personalists tend to focus more on the act itself and its contribution to the attainment of certain human personal and communal goals and goods revealed by nature. Thus, the former approach tends to be more tolerant and approving, whereas the latter tends to be less tolerant and accepting. McCartney thinks that the challenge in Catholic moral theology is to reconcile these approaches, preserving what is of enduring value in each. Further, he favors less discussion about specific acts in sexual ethics, and calls for an examination for them in the context of intimacy, spirituality, and interpersonal growth. This would entail talk about the relation of love and physical embodiment, human love as a symbol of divine love, the meaning of friendship and commitment, and improvement of the quality of family life.

Robert Springer's theologically-oriented analysis of transsexualism and surgical sex reassignment follows. Thirty years ago the author notes, theological ethicists rejected transsexual surgery as "a grave mutilation of the human body." More recently, the trend has been to approve of transsexual surgery for true hermaphrodites and in situations, which most would consider hypothetical, in which a person could be determined by some "objective" standard to be of the opposite gender than he or she is anatomically and physiologically. Springer believes that such a situation is possible. Current research indicates that gender identity is determined primarily by parental sex assignment and rearing, not anatomy and physiology. What, then, is the biblical and theological "ought" in such cases? Springer dismisses the quietistic notion, which has dominated thinking in this area, that humans ought not to tamper with God's creation. According to the author, God's granting humanity the dominion over creation implies a delegation of responsibility to actively manage creation to the best of our ability. The biblical mandate is that we "choose life," that is, that we use technology to promote human life and human wholeness. If more conservative treatments fail, and if transsexual surgery will help a person develop this sense of wholeness, then it is consistent, in Springer's opinion, with the biblical mandate that sex reassignment surgery be performed.

The third section ends with Ronald Green's examination of the hypothesis that most religious beliefs about sexual ethics, as well as many of the Western religious beliefs not specifically about sexual ethics, but underlying the latter, may be of no value or irrelevant. He examines this hypothesis by considering how cultural and scientific developments have challenged four traditional religious beliefs regarding sexuality: (1) that sexual conduct is morally and religiously significant; (2) that sexual relations are proper only in a lifelong personal relationship with another person; (3) that heterosexuality is morally and religiously normative; and (4) that biological gender is an appropriate basis for categorizing and assigning social roles to persons. In addition to examining the biblical roots of these beliefs and the secular sources of the challenges that confront them, Green discusses the sexual theologies of Karl Barth, Helmut Thielicke, and James Nelson. Green thinks that the biblical sources of beliefs about human sexuality are largely irrelevant in light of more modern understandings. And he charges that modern theologians remain too bound to irrelevant biblical texts, drawing their attention away from modern research on sexual topics. Only by revising the priorities that they assign to the sources available to them, according to Green, will modern sexual theologies become more adequate.

The authors show, in their individual contributions and in the collection as a whole, that the sources of dissent about matters of sexual morality in many instances are of ancient origins, involving differing fundamental visions of the good of and for humanity. Whether grounded in secular or religious traditions, diverse, often conflicting, and perhaps irreconcilable judgments abound regarding norms for human sexual expression. Establishing one or more vision today of what is and is not morally licit about human sexuality seems beyond reach at present. This means, of course, that the involvement of medicine and physicians with people about their sexuality will continue to be controversial. Like sexuality, medicine and medical practice are capable of serving a variety of ends in responding to the desires and complaints of persons. These varied responses illustrate the dynamic character of contemporary understandings of the purposes of medicine and the role of physicians in society. Additionally, and as importantly, the facility with which medicine and physicians can embody and embrace pluralistic values is revealed. Thus debate about issues in sexuality and medicine goes on.

These observations, however, are not meant to imply that scholars, medical professionals, and people in the role of patient ought to be resigned to an endless dispute with nothing to be gained from continued reasoned investigations and discussions. On the contrary, the current struggle over sexuality and medicine ought to be seen as an opportunity to engage in the challenging work of examining again the received wisdom about these matters of contention. This activity may produce new insights and better answers to perplexing medico-moral questions. What was held to be true may be affirmed. The received tradition might, as well, require partial or complete revision when examined in the light of the results of contemporary research in the medical and social sciences, philosophy, and religion. A better understanding of points of disagreement and agreement could be a valuable result of these investigations and dialogue, even if consensus is not reached. This sort of open-minded study and conversation could help physicians and patients to appreciate the pluralistic character of the moral order and the rich texture of human sexuality. Further, by directing the same sort of critical light on medical presuppositions, understandings, and values regarding human sexuality, progress might be made toward a more comprehensive understanding of the legitimate tasks of medicine and physicians with respect to human sexuality. In the end, perhaps everyone will have a better sense of where physicians are properly authorized to evaluate, control, or alter human sexual function or expression.

Many people have made this volume possible. The contributors completed their assignments in a timely and competent manner rendering the work of the editor less onerous. Each one should know of my appreciation for their charitable work and enduring commitment to this project. The series' editors, H. Tristram Engelhardt, Jr. and Stuart F. Spicker, were helpful throughout the process of putting these essays in published form. Their counsel, patience, advice, and encouragement are appreciated. Jay Jones, my research assistant, expertly reviewed manuscripts for errors and omissions. His thoroughness kept some contributors from being embarrassed and my work as editor much easier. Susan M. Engelhardt proofread the manuscript, helping to assure that the contributors' work is presented accurately and in consistent form. Lolita Cannon, my secretary, performed various assignments related to this project with characteristic skill and good humor. All of these people have made this book possible. I thank them all, including the staff at D. Reidel Publishing Company.

Institute of Religion, and
Center for Ethics, Medicine and Public Issues,
Baylor College of Medicine,
Houston, Texas, U.S.A.

EARL E. SHELP

BIBLIOGRAPHY

1. Shelp, Earl E. (ed.): 1987, *Sexuality and Medicine, I: Conceptual Roots*, D. Reidel Publishing Co., Dordrecht and Boston.

SECTION I

REPRODUCTION, MEDICINE, AND MORALS

MARY ANN GARDELL

SEXUAL ETHICS: SOME PERSPECTIVES FROM THE HISTORY OF PHILOSOPHY

INVESTIGATING THE INTERPLAY BETWEEN SEXUALITY AND ETHICS

Human life is a complex and rich endeavor, of which sexuality is but one element, albeit centrally important. As an element of a complex, the significance of sexuality must be judged in terms of its place in the constellation of human affairs. Moreover, concepts of sexuality have an evolving character. This is in part due to changes in our views of sex, which in turn depend on diverse evaluations and understandings of the world. These change and develop with changes in philosophical perspectives. Consequently, the significance of sexuality depends on its position with reference to a wide range of human values. Conversely, the position one gives to sexuality will determine the significance assigned to other endeavors we engage in. This essay investigates the significance philosophers have assigned to sexuality, particularly as it contributes to the good or moral life of individuals: How have philosophical traditions understood the interplay among issues in sexuality, ethics, and the lives of persons? Is there an ethics unique to sex? Are canons of moral probity regarding sex to be found in some concrete normative understanding of sexuality? Or, might they be found in that which is intrinsic to morality itself? If neither account is sustainable, what then can be said about the character of sexual ethics? Our answers to these questions are the culmination of two millenia of reflections.

In investigating the character of sexual ethics, one confronts major ambiguities in the terms "sexuality" and "ethics". "Sexuality" may refer generally to a wide range of considerations associated with ourselves as biological and gender-conditioned individuals. These considerations are expressed in the descriptive, explanatory, evaluative, and social performative roles of language[1] regarding being sexual and

Earl E. Shelp (ed.), Sexuality and Medicine, Vol. II, 3–15.

having sex. These considerations, though they may be distinguished, inseparably interplay in fashioning the complex character of sexuality.

Consider, for example, the ways in which we understand maleness. A recent *New England Journal of Medicine* article, entitled 'The Search for the Ultimate Cause of Maleness' [18], argues that certain biological features determine the character of maleness. In such accounts, descriptive features are understood against the background of explanatory theories that indicate the relevant attributes associated with maleness. Furthermore, concepts of maleness involve value judgments, considerations that turn on the goodness or desirability of being male. Concepts of sexuality also serve as directives or actions warrants in particular communities. These warrants guide male behavior within social institutions, such as courtship, marriage, and family structure. In these ways, concepts of sexuality are inextricably theory-ladened, value-infected, and socio-historically conditioned.

The roles of sexuality are complex and include reproduction, social or institutional stability, and recreation ([7], p. 51). Though some hold that reproduction, or the procreation of children, is the cardinal goal of sexuality (e.g., [16], [15]), others claim that social stability, or the fostering of stable relationships among friends and marriage partners provides an equally, if not more important, goal ([29], [27], [17]). Then again, others defend the view that sexuality plays an important recreational role apart from commitments either to reproduction or significant social bonds [28]. In short, the social roles of sexuality reflect what individuals hold to be significant about the character of being sexual and of having sex.

Our understandings of "ethics" are as equivocal as our understandings of sexuality. Generally put, ethics as a philosophical discipline is devoted to the study of the moral values, virtues, and the principles of action that bind individuals and provide common grounds for blame and praise and for delineating the character of the good life. In addition, there are numerous other senses of ethics or morals: the rules of conduct as determined by the customs of a people, the etiquette of a profession, the rules of law, and the teachings of a religious doctrine. All of these contrast in different ways with the results of rational analysis, including both plausible and conclusive accounts of proper moral conduct ([6], p. 54). In short, "ethics" has come to be associated with particular moral views of a culture, profession, legal body, religion, or with the community of rational agents as determining what should be generally endorsed or normatively correct. It is this last sense of ethics that has been the focus of philosophical thought.

This philosophical sense of ethics can itself be resolved into two major senses. To begin with, ethics attempts to provide a basis for the resolution of interpersonal conflicts on grounds other than force or coercion. The cardinal principle here is respect for persons, a notion indebted to Kant [13] and developed further in Post-Enlightenment reflections ([24], [6]). This tier of moral life rests on a "content-poor" secular foundation that attempts to span numerous divergent moral communities through conclusive arguments regarding the character of reason. This first tier of the moral life may be contrasted with a second where appeal is made not to the notion of persons as free moral agents but instead to a particular hierarchy of achievable goods. In this second tier, what is provided are, at best, plausible arguments, and not conclusive ones.

Pace procrustes, Western philosophical thought regarding the character of sexual ethics can be ordered around four major positions. First, there are those who suggest that little, if anything, of unique significance can be assigned to the nature of sexual ethics. Philosophers of the Pre-Christian Era, such as Plato (428–348 B.C.) and Aristotle (384–322 B.C.), offer a foundation for this view. Second, there are those who argue that the basis of sexual ethics is to be found in some concrete understanding of sexuality. This position was developed by traditional Christian thinkers such as St. Augustine (354–430) and St. Thomas Aquinas (1224–1274). Third, there are those who argue that the character of sexual ethics may be understood in terms of the application of basic ethical principles to issues regarding sexuality. This view emerges during the Enlightenment Era, particularly with Immanuel Kant (1724–1804). However, as we will see, Kant fails to realize this goal. A more recent attempt in this view has been provided by H. Tristram Engelhardt, Jr. [7].

COMPETING VISIONS OF SEXUAL ETHICS

Ancient Greek and Roman philosophers provide a backdrop to appreciate Pre-Christian views on the morality of sexuality. One of the major figures, Plato, defends the view that reason should never be obtunded either by pleasure or by corporeal or bodily matters. Consider the dialogue between Socrates and Callicles:

Socrates.	Then the house in which order and regularity prevail is good; that in which there is disorder, evil?
Callicles.	Yes. . . . .
Socrates.	And the same may be said of the human body?
Callicles.	Yes (*Gorgias*, 503, [25]. Vol. I, p. 565).

Elsewhere Socrates assigns pleasure a place among the last and lowest of the goods:

Socrates.	The claims both of pleasure and mind to be the absolute good have been entirely disproven in this argument, because they are both wanting in self-sufficiency and also in adequacy and perfection.
Protarchus.	Most true.
Socrates.	But, though they must both reign in favour of another, mind is ten thousand times nearer and more akin to the nature of conqueror than pleasure.
Protarchus.	Certainly.
Socrates.	And, according to the judgment which has now been given, pleasure will rank fifth.
Protarchus.	True (*Philebus*, 67, [25], Vol. II, p. 403).

With regard to sexual pleasure, the Athenian condemns homosexual practices on grounds that they are "contrary to nature" in that they were spawned originally from "unbridled lust" (*Phaedo*, 65, [25], Vol. I, p. 448). In short, Plato suggests that sexual pleasure, and the activities accompanying it, are distant from, if not in tension with, the moral life of reason and control.

The theme of rational control is developed further in Stoic philosophical thought, particularly in that of Epictetus. As Epictetus warns us, "When you yield to carnal passion you must take account not only of this one defeat, but of the fact that you fed your incontinence and strengthened it" (*The Discourse and Manual*, Bk. 2, Ch. 18, p. 108, [8], p. 208). This view is reflected as well in the social structures and practices of early Greco-Roman times. Recall here that marriages were early and for the most part under strict parental control. In addition, class barriers, the stability of communities, and the absence of social mobility severely restricted the choice of sexual and marriage partners ([21], pp. 254–255).

It would be misleading to suggest that Plato offers a univocal account of sexuality and of its role in the good life. A second major interpretation is in fact available. Plato assigns Eros a role in the achievement of life's greatest good, namely, the contemplation of pure beauty. As Plato tells us, "the love, more especially, which is concerned with the good, and which is perfected in company with temperance and justice, whether among gods or men, has the greatest power, and is the source of all our happiness and harmony, and makes us friends with the gods who are above us, and with one another" (*Symposium*, 188, [25], Vol. I, p. 315). Furthermore, ". . . through Love all the intercourse and converse of god with man, whether awake or asleep is carried on" (*Symposium*, 203, [25], Vol. I, p. 328). In this way, man may come to possess the good life of the beautiful and of divine transcendence. Sexuality is transformed into an aesthetic ideal. In fact, love between two males is commented on in a positive fashion:

> Those who are inspired by this love [Aphrodite] turn to the male, and delight in him who is the more valiant and intelligent nature; any one may recognise the pure enthusiasts in the very character of their attachments. For they love not boys, but intelligent beings whose reason is beginning to be developed, at the time at which their beards begin to grow (*Symposium*, 181, [25], Vol. I, p. 309).

In short, sexual pleasure, sexual intercourse, and love are esteemed, if not indeed romanticized.

Plato may be seen to offer two major interpretations of sexuality. Sex is viewed, on the one hand, as alien to the life of reason and, on the other, as contributing in important ways to the achievement of a utopian existence. This tension between competing views of sexuality is seen as well in the writings of Aristotle. The good life, for Aristotle, consists of the contemplation of the life of one's friend. This activity, he tells us, has room neither for the incorporation of sexual activity nor for the nurturing power of erotic love. This is the case in part because pleasure in general, and the joys of sex in particular, cannot be immediate objects of moral judgment in the good life. They are subjective in character. Only objective phenomena, such as actions, may be judged as morally praiseworthy or blameworthy. Aristotle does admit later, however, that acts and pleasure are inseparably joined, since "when both object and perceiver are of the best there will be always pleasure" [*Nichomachean Ethics*, 1174b, [2], p. 1099]. Accordingly, a judgment of the act carries with it a judgment on the attendant pleasure. "The pleasure proper to a worthy activity is good and that proper to an unworthy activity bad" [*Nichomachean Ethics*, 1175b, [2], p. 1011]. Nevertheless, the Aristotelian distinction between actions as objective and pleasures as subjective serves to remove pleasure, passions, and desires from cardinal roles in the good or moral life.

On the other hand, Aristotle expresses an appreciation of the role pleasure plays in the cultivation of the good life. Friendships between a husband and wife depend, according to Aristotle, on the achievement of the *virtuous life* shared by two individuals, of *stability* in the formation of a household, and of *pleasure* in the act of generation. In short, the marital friendship, with its threefold basis, involves the pleasures of coition (*Nichomachean Ethics*, 1162a, [2], p. 1073). Elsewhere Aristotle joins Plato in claiming that the highest form of friendship is that which is held between two men (*Nichomachean Ethics*, 1162b, 1163b, [2], pp. 1073–1076).

In summary, Plato and Aristotle offer an appreciation of sexuality that is both alien to as well as integral to the good or virtuous life. The former view presumes the world is a composite of finite entities functioning within universal laws of nature. The essences of these entities are discoverable by reason, i.e., a special faculty of *nous*, in such a way that necessary and commensurately universal first principles may be advanced. Those objects not amenable to rational scrutiny fall outside of the scope of philosophers' concerns. Thus, one will be able to find in neither Plato nor Aristotle, for example, discussions of the normative character of masturbation, sodomy, or lust.

In contrast, early Christians, such as St. Augustine and St. Thomas Aquinas, offer extended reflections regarding the character of sexual ethics. The goal of the good Christian life is, according to St. Augustine, rational and in accord with the Divine Will. That which deviates from the rational is against the Divine Order, and hence evil or sinful. However, one can reorder evil tendencies according to reason, thus integrating them into right and proper order. Sexual pleasure becomes ordered, for example, when it is brought in line with the goal of procreation. "What food is to the health of man, intercourse is to the health of the species" (*The Good of Marriage*, 16.18, [23], p. 127). Likewise, "The procreation of children is the first and natural and lawful reason for marriage" (*Adulterous Marriages*, 2.12.12, [23], p. 138). This suggests that in the end intercourse acquires an unqualified positive role in the marital bond. Coition is always, according to St. Augustine, tainted with the evil or sin of pleasure, or concupiscence. This is the case because man's generative parts will not obey reason (*Marriage and Concupiscence*, 1.6.7, [23], p. 140). In the act of copulation, disobedience is always manifest: "Although conjugal copulation to generate offspring is not itself a sin (because the good will of the mind controls the pleasure of the body instead of following pleasure's lead, and the human judgment is not subjugated to sin) yet in the use of the generative act the wound of sin is justly present" (*Marriage and Concupiscence*, 1.12.13, [23], p. 134). This may be to suggest that the loss of rational integrity that accompanies sexual passion is passed on from one generation to another. This may in fact be through the very character of procreation itself ([23], p. 132). As St. Augustine reminds us, if not for sin, "marriage, worthy of the happiness of Paradise, should have had desirable fruit without the shame of lust" (*City of God*, Ch. XXIII, [3], p. 268). For if man had only remained innocent and obedient in Paradise, intercourse would be known to us as follows: "the man, then, would have sown the seed, and the woman received it, as need required, the generative organs being moved by the will, not excited by lust" (*City of God*, Ch. XXIV, [3], pp. 369–370). St. Augustine characterizes marriage as good, sexual lust, if not pleasure, as an evil, and marital intercourse, in the Fallen State, as a mixture of the two.

In line with St. Augustine, St. Thomas Aquinas develops what may be referred to as a "procreative sexual ethics". St. Thomas argues for the goodness of sexual pleasure insofar as it is ordered according to rational goals. For St. Thomas, all things, including our bodies, are endowed with intrinsic goals. Designed by Nature, and ultimately by God, these goals are natural facts discoverable by reason. For example, Nature – or God – assigns to sexual intercourse the goal of procreation.[2] Accordingly, sexual intercourse, and the pleasures accompanying it, are to be engaged in only to the extent that offspring are intended or in order to avoid adultery. These norms are natural because "[r]eason discovers good purposes through the very facts of the function of sexual organs" (*Summa Contra Gentiles*, III, 122, 4 and 5, [23], p. 246), and their function is to emit "human seed, in which man is in potentiality" (*Evil* 15.2, [23], p. 244). And "[t]he disordered emission of seed is contrary to the good of nature, which is the conservation of the species" (*Summa Contra Gentiles*, III, 122, [23], p. 245). Furthermore, the conjugal act, with the intent to procreate or to fulfill the "marital debt", must occur in the context of a heterosexual, marital relation. To reason or to act otherwise is to violate fundamental laws of Nature, thereby engaging in crimes against reason and ultimately against God. "Just as the ordering of right reason proceeds from man, so the order of Nature is from God himself; wherefore in sins contrary to nature, whereby the very order of nature is violated, an injury is done to God, the ordainer of Nature" (*Summa Theologica* II–II.154.12, obj. 1, [23], p. 244). In this way, all modes of sexual activity deviating from norms of Natural Design are unnatural, and hence immoral. This includes any attack on human life, even human life potentially present in the human seed (i.e., sperm) or any impairment of the normal process of preserving the human species. Non-procreative sexual relations in general, and sodomy, homosexuality,[3] and masturbation in particular are thereby proscribed on moral grounds.

Unlike St. Augustine, St. Thomas offers a more positive view of the relation between sexual intercourse and the marital bond. Marital coitus may in fact be without sin (*Summa Theologiae* II–II.153.2, [23], p. 256) when concupiscence and intercourse are excused, or compensated for, by the good of procreation. Moreover, St. Thomas may be seen to have developed a theory of love within which sexual union may be an aid to interpersonal love (*Summa Theologiae* II–II, 26, 11, [23], p. 256). Whether St. Thomas actually breaks with St. Augustine's theory of procreative sex and fully justifies marital intercourse as an expression of the good of fidelity is an issue that receives attention long past the thirteenth century ([10], pp. 9–11).

Fifteenth through eighteenth century reflections call into question traditional presumptions regarding the character of sexuality. Those writing during the Protestant Revolution, such as Martin Luther [19] (1483–1546) and John Calvin [4] (1509–1664), each challenge the view that Natural Design, as ordained by God, discloses the morality of particular sexual attitudes and behaviors. They claim that sexual pleasure and its activities may be understood as simple facts of life, providing a basis for mutual bonds between husbands and wives, who may wish to bring offspring into this world.

The less one is persuaded that God exists or that one can discern, and in turn violate, Natural Design, the more one will seek alternative grounds for resolving conflicts regarding the morality of sex. Enlightenment philosophers such as Immanuel Kant (1724–1804) advance alternative grounds for sexual ethics derived from the character of reason itself. Kant views the Christian's attempt to secure content for sexual morality from Nature as futile. In an attempt to give sexual morality rational grounding, he recaptures in categorical reasoning Christian arguments regarding the morality of particular sexual attitudes and behaviors. He argues that if participants in sexual activities use each other as means rather than as ends to attain individual sexual gratification, then their relationship is one of mutual masturbation. As he says, "Taken by itself it is a degradation of human nature; for as soon as a person becomes an Object of appetite for another, all motives of moral relationship cease to function, because as an Object of appetite for another a person becomes a thing and can be treated and used as such by everyone" ([14], p. 163). For Kant, treating another, even oneself ([14], p. 170), as a means is immoral. One recalls here Kant's first formulation of the categorical imperative: "Act only according to that maxim by which you can at the same time will that it should become a universal law" ([13], p. 39).[4] To will that all individuals are to be treated as means is to contradict the very notion of the moral community: "Act so that you treat humanity, whether in your own person or in that of another, always as an end and never as means only" ([13], p. 47). However, the marital bond is seen by Kant to unify two individuals so that neither uses the other merely as a means. "Matrimony is an agreement between two persons by which they grant each other equal reciprocal rights, each of them undertaking to surrender the whole of their person to the other with a complete right of disposal over it. We can now apprehend by reason how a *commercium sexuale* is possible without degrading humanity and breaking the moral laws ([14], p. 167). In this way, marriage transforms an otherwise manipulative masturbatory relationship into one that is essentially altruistic in character.

On Kant's view, reason delivers sufficient content for a generally persuasive account of sexual morality. This is possible, however, through Kant's view of autonomy, which excludes any role for passion or lust in determining the moral will. To act on grounds of passion or lust is to act heteronomously, or immorally, since one is not acting as a moral agent as such. To have sexual intercourse for pleasure is thus immoral for Kant as it is for St. Augustine. In this way, Kant vindicates Christian morality without appeal to Natural Design and its Deity. Kant's argument, it should be noted, depends on freedom and rationality being goals of moral action, not conditions of moral action. If they are only the latter, then one may in many circumstances freely and rationally, and indeed morally, decide to engage in sexual activity for sexual pleasure alone.

David Hume (1711–1776) attempts to ground sexual morality neither in an understanding of the nature of sexuality, as the Christians saw it, nor in some concrete view of rationality, as does Kant. Rather, Hume gestures toward a different direction of argument. On the one hand, Hume defends the idea of the traditional Western marriage (i.e., "an engagement entered into by mutual consent and has for its end the propagation of the species" ([12], p. 232). He argues that polygamy and divorce are inimical to these ends. He concludes that "the exclusion of polygamy and divorce sufficiently recommend our present European practices with regard to marriage" ([12], pp. 231–239). On the other hand, however, Hume makes a mockery of many traditional arguments. As he says, reasonings "which cost so much pains to philosophers" and which "are often formed by the world naturally and without reflection" are often plagued "with difficulties which seem unsurmountable in theory" but which "are easily got over in practice" ([11], p. 124). Recounting the argument that men should be courageous and women should be chaste, he concludes that "bachelors, however debauched, cannot choose but be shocked with any instance of lewdness or impudence in women" ([11], p. 124). One can imagine here Hume having just seduced a young girl and then reproving her on her loss of virtue as he courageously strides out the door of her boudoir. Hume's comical treatment of his predecessors' visions of sexual morality serves to suggest the futility of discovering in Natural Design or in reason alone the concrete nature of sexual ethics. His approach to these questions can be interpreted as a *reductio ad absurdum* of traditional moral doctrine.

Nineteenth and twentieth century philosophical thought[5] challenges attempts, such as Kant's, to provide a concrete sexual ethics. The more one recognizes the inability of reason to render conclusive moral arguments regarding the concrete character of the good moral life, the more one will be compelled to search for alternative grounds for resolving conflicts regarding the morality of particular sexual activities. Despite his novel attack on traditional versions of sexual morality, Hume fails to develop at any great length an alternative approach to understanding sexual morality. One might conclude, then, that a generally defensible view of sexual ethics is unavailable and that any attempt to discover content for sexual morality is futile. A nihilism regarding sexual ethics appears very likely. This intellectual conclusion was supported, at least psychologically, by the descriptive relativism of cultural anthropologists.

Yet, one might claim along with H. Tristram Engelhardt, Jr., that there is available a generally defensible understanding of ethics that can be applied to issues regarding sexuality. Engelhardt argues that concrete canons for moral guidelines, such as those governing sexual attitudes and practices, offer at best plausible arguments. They do not provide conclusive accounts of the moral significance of sexuality. However, general moral principles can be secured and applied to sexuality such as canons of promise-keeping, truth-telling, and mutual respect. A major implication of this view is that it is futile to expect either religious or philosophical traditions to supply generally justifiable concrete goals for human sexuality. As Engelhardt puts it:

> The moral character of monogamous, polygamous, polyandrous, and even more complex sexual alliances, heterosexual and homosexual alike, would seem to depend on the character of the commitments made and the likely benefits and harms to ensue, without any claim having been advanced about moral issues unique to sexuality ([7], p. 53).

Here, one is reminded of the ancient Greco-Roman vision of sexuality as tolerant of differences among particular sexual practices insofar as certain principles (e.g., social stability) were not violated. The principles that may be generally defended in contemporary society have, of course, changed. Western culture recognizes the central importance of respect for persons, known also as the principle of autonomy, in the moral life. Putting differences aside, however, contemporary philosophical reflections in concert with ancient ones remind us of the futility of discovering an ethics unique to sexuality.

SEXUALITY, ETHICS, AND THE LIVES OF INDIVIDUALS

This overview suggests distinct, yet overlapping, philosophical traditions regarding the morality of sexual attitudes and practices. Post-Enlightenment thinking reminds us of the unavailability of a unique sexual ethics. There may be, as Engelhardt argues, only general principles such as promise-keeping, truth-telling, and mutual respect, that may guide the search for proper sexual attitudes and practices. For example, strong cases of moral degradation in sexual activity (e.g., rape) will only be those that involve the use of others without their consent.[6] Accordingly, the content of a sexual ethics must be acknowledged as heterogeneous and complex in character, fashioned by a wide range of claims, interests, desires, and goals that influence the ways in which we describe, explain, and evaluate sexual behavior. This accommodation to a pluralism of moral intuitions regarding sexuality is not to suggest that particular ethics must be abandoned. Rather, particular convictions are encouraged as private commitments offering ways in which interpersonal meaning can be realized for a life of moral content. In particular contexts, these convictions may in fact be considered morally binding in the strong sense.[7]

Center for Ethics, Medicine, and Public Issues,
Baylor College of Medicine,
Houston, Texas, U.S.A.

NOTES

[1] Descriptive considerations indicate the structural features we associate with a particular object of concern. Explanatory ones render intelligible the ways in which relevant factors and features interplay in structuring the world around us. Evaluative considerations indicate the significance we assign to our observations and endeavors. And social performative ones indicate the ways in which concepts function as action directives, warrants, or imperatives. Further extrapolation on these points is found in Engelhardt ([16], Ch. 5).
[2] In the words of Genesis (8:17), man is to "increase and multiply and fill the earth".
[3] One recalls here the passage from Leviticus (18:22–23): "thou shall not lie with mankind as with womankind . . .", and from I Corinthians 6:10: "Nor the effeminate, nor liers with mankind . . . shall possess the Kingdom of God".
[4] Kant would not have argued that there are no moral rules against sexual activities as long as the freedom of others is not violated.
[5] See, e.g., Mill [20], Nietzsche [22], Grene [9], Nagel [21], Solomon [28], and Symons [29].
[6] Ruddick [27] does not subscribe to this position. She holds, rather, that expressions of extreme domination (e.g., sadism) are to be excluded as well.
[7] Consider, for example, the Encyclical *Humanae Vitae*, issued July 29, 1968, by Pope Paul VI, where it is stated that "the direct interruption of the generative process already begun, and, above all, directly willed and procured abortion, even if for therapeutic reasons, are to be absolutely excluded as licit means of regulating birth" (*Humanae Vitae* [26], p. 138).

BIBLIOGRAPHY

1. Aquinas, St. Thomas: 1945, *Summa Theologiae*, trans. A. C. Pegis, Random House, New York.
2. Aristotle: 1941, *The Basic Works of Aristotle*, ed. R. McKeon, Random House, New York.
3. Augustine, Saint: 1948, *Basic Writings of Saint Augustine*, ed. W. J. Oates, Random House, New York, Vols. I and II.
4. Calvin, J.: 1981, *Commentary on Genesis*, in S. Dunn (ed.), *The Best of John Calvin*, Baker Book, New York, pp. 49–63.

5. Engelhardt, H. T., Jr.: 1974, 'The Disease of Masturbation: Values and the Concepts of Disease', *Bulletin of the History of Medicine* **48**, 234–248.
6. Engelhardt, H. T., Jr.: 1986, *The Foundations of Bioethics*, Oxford University Press, New York.
7. Engelhardt, H. T., Jr.: 1987, 'Having Sex and Making Love: The Search for Morality in Eros', in this volume, pp. 51–66.
8. Epictetus: 1916, *The Discourse and the Manual*, trans. P. E. Matheson, Clarendon Press, Oxford.
9. Grene, M.: 1979, 'Comments on Pellegrino's "Anatomy of Clinical Judgments" ', in H. T. Engelhardt, Jr. *et al.* (eds), *Clinical Judgment: A Critical Appraisal*, D. Reidel Publishing Company, Dordrecht, pp. 195–197.
10. Grisez, G. G.: 1966, 'Marriage: Reflections Based on St. Thomas and Vatican Council II', *Catholic Mind* (June), 4–19.
11. Hume, D.: 1951, 'Of Chastity and Modesty', in F. Watkins (ed.), *Hume: Theory of Politics*, Thomas Nelson and Sons, Ltd., Toronto, pp. 121–125.
12. Hume, D.: 1875, 'Of Polygamy and Divorces', in T. H. Grene and T. H. Grose (eds.), *Essays: Moral, Political, and Literary*, Longsman and Green, London, pp. 231–239.
13. Kant, I.: 1959, *Foundations of the Metaphysics of Morals*, trans. L. W. Beck, Library of Liberal Arts, Indianapolis.
14. Kant, I.: 1963, 'Duties Towards the Body in Respect of Sexual Impulse', in I. Kant, *Lectures on Ethics*, trans. L. Infield, Harper and Row, New York, pp. 162–168.
15. Kelly, G.: 1958, *Medico-Moral Problems*, Catholic Hospital Association, St. Louis.
16. Kenny, J. P.: 1962, *Principles of Medical Ethics*, Newman Press, Westminster.
17. Ketchum, S. A.: 1980, 'The Good, the Bad, and the Perverted: Sexual Paradigms Revisited', in A. Soble (ed.), *Philosophy of Sex: Contemporary Readings*, Littlefield, Adams, and Co., Totawa, pp. 139–157.
18. Kidd, K. K.: 1985, 'The Search for the Ultimate Cause of Maleness', *New England Journal of Medicine* **313**, 260–261.
19. Luther, M.: 1983, *A Sermon on the Estate of Marriage*, in J. N. Lenker (ed.), *Sermons of Martin Luther*, Vol. 3, Baker Book, New York, pp. 45–56.
20. Mill, J. S.: 1966, 'The Subjection of Women' (1869), reprinted in *Three Essays by J. S. Mill*, Oxford University Press, London, p. 427.
21. Nagel, T.: 1969, 'Sexual Paradigms', *Journal of Philosophy* **66**, 5–17.
22. Nietzsche, F. W.: 1970, *Beyond Good and Evil* #168, in *The Philosophy of Nietzsche*, Random House, New York, pp. 465–466.
23. Noonan, J. T., Jr.: 1966, *Contraception: A History of Its Treatment by the Catholic Theologians and Canonists*, Harvard University Press, Cambridge.
24. Nozick, R.: 1975, *Anarchy, State, and Utopia*, Basic Books, New York.
25. Plato: 1982, *The Dialogues of Plato*, trans. B. Jowett, 14th edition, Random House, New York.
26. Paul VI, Pope: 1968, *Humanae Vitae*, in R. Baker and F. Elliston (ed.), *Philosophy and Sex*, Prometheus Books, Buffalo, pp. 131–149.
27. Ruddick, S.: 1975, 'Better Sex', in R. Baker and F. Elliston (ed.), *Philosophy and Sex*, Prometheus Books, Buffalo, pp. 83–104.
28. Solomon, R.: 1974, 'Sexual Paradigms', *Journal of Philosophy* **71**, 336–345.
29. Symons, D.: 1979, *The Evolution of Human Sexuality*, Oxford University Press, New York.

SARA ANN KETCHUM

MEDICINE AND THE CONTROL OF REPRODUCTION

The development of technology makes it possible for us to make choices about reproduction which were not in our power a short time ago. New technologies for controlling fertility, for detecting genetic abnormalities, and for detecting fetal defects in utero raise the possibility of making choices about the quantity and quality of the next generation. This possibility forces on us questions not only about what decisions ought to be made but also about who should make them. This paper will examine the moral claims of prospective parents, health professionals, future generations, and the society in general to have a say or have their interests considered in such decisions.

Regardless of what conclusions we come to about who has a right to a say in these decisions, physicians and health professionals will be faced with making or influencing decisions about these matters. Since it is the technology that gives us (as a society or a species) the power to make such decisions effective, those who have the capacity to use that technology will have a measure of control over those decisions. Physicians will be the ones immediately making the decisions, since they will be the ones performing the actions. For example, physicians perform the tests used to detect fetal abnormalities in utero and would perform either an abortion or corrective surgery if one of these options is chosen; and, given the present structure of the health professions, those procedures performed by nonphysicians (for example, by nurses or technicians) will be done under the supervision of physicians.

Given this fact, it might be tempting to dismiss the question with the formula of leaving the decision to the patient in consultation with her physician, as does the Supreme Court in *Roe v. Wade*. However, from a moral point of view, this is unsatisfactory for reasons that go beyond the general problem – that an answer to the question of who should make a decision does not give us any guidance about what should be the content of that decision. Reproductive decisions raise moral problems not raised by other medical decisions.

Earl E. Shelp (ed.), Sexuality and Medicine, Vol. II, 17–37.

I will start with the democratic principle that anyone who has a stake in a decision – that is, anyone whose rights or interests will be affected by the outcome of the decision – has at least a *prima facie* right to have a say in that decision. Normally, in medical decisions we assume that the stake that the patient has in the result of the decision is so much greater than that of anyone else who might be affected that it is appropriate to leave the decision entirely up to the patient, within the limits of technical feasibilities set by the physician. Some of the exceptions to this policy occur with contagious diseases, organ transplants, and use of scarce technology in the form of machines such as kidney dialysis machines. In these cases, issues of justice (in distributing scarce technology) or of the significant interests of others counterbalance concern for the autonomy of the patient. But, in the absence of such considerations, it is appropriate to assume that medical decisions are private (that is, that they directly affect only the patient) and should be left to the autonomous decision of the patient.[1]

However, reproductive decisions have a special set of problems that are raised by the fact that the future existence or nonexistence of some persons will be a part of the content of the decision. Decisions not to reproduce appear to be private; they do not directly affect any third party. But, a decision to reproduce creates a third party who is crucially affected and who did not consent to the decision. Not only did this third party not consent, but she or he is not available for consultation. The nonexistence of this crucial third party at the time of the decision and our interest (as human beings and as members of a given culture) in the future existence of a next generation raise several interesting moral problems and suggest an argument for a greater social control of reproductive decisions than of nonreproductive medical decisions. However, I will argue that, despite these problems, reproductive decisions should typically be treated as private.

RIGHTS OF PROSPECTIVE PERSONS

The argument that negative reproductive decisions (decisions not to reproduce) are private while positive reproductive decisions (decisions to reproduce) are public suggests that we are only morally required to count those people who will exist if the reproductive decision is positive but not those who would have existed if the decision had been positive. But if the child that I will have if I decide to have a child has a stake in my decision, does not the child I would have had if I had not decided not to have a child have an even greater stake in my decision? Before I proceed with this question, I want to introduce some definitions. I shall use the term "prospective person" for those persons whose existence or nonexistence is dependent on the decision or decisions under consideration,[2] and 'counterfactual person' for those persons who would have existed if someone had made a reproductive decision that they in fact did not make. Note that counterfactual persons are fictional; they are persons who would have existed if conditions had been different, but they do not and will not exist. Prospective persons are (will become) either counterfactual persons or (actual) future persons.

Although we do have obligations to future persons, there are good reasons to deny that we have obligations to counterfactual persons, or that we have an obligation to prospective persons to make them actual. Suppose I decide not to have any children at all, and decide on a surgical sterilization. Apart from natural law arguments against any artificial interference in the process of reproduction, most would assume that such a decision is not immoral. Indeed, even those who adhere to the natural law arguments would find the decision unobjectionable as long as the method used for avoiding reproduction is celibacy rather than surgical sterilization. However, if we were morally required to take into consideration the interest in being produced of prospective persons, then the person who remains childless would have denied existence to – and thereby at least *prima facie* wronged – a virtually infinite number of counterfactual persons. This would follow if we were to grant rights to all prospective persons. Since the number of such persons one could actually produce in a lifetime is biologically limited, one may have obligations only to produce as many as possible. Thus, if particular prospective persons have rights, we have to violate (or override) the rights of some of them in order to insure the existence of others. Moreover, if we decide to produce as many persons as we can, we increase the ecological burden on the earth and seriously reduce the quality of life for those persons who do and will exist. R. M. Hare, who wants to adopt something like the position that we have an obligation to prospective or potential persons to bring them into existence, suggests that if there is some maximally beneficial number of persons that the world could support or that I could take care of, we need not be committed to the position that I have done wrong to every counterfactual person I could have produced, although I have wronged counterfactual persons whose lives would have been happy without significantly affecting the lives of other actual and possible persons ([7], p. 218). Although Hare's argument saves us from the burdensome obligations of continuous reproduction, it still leaves us with significantly counterintuitive consequences: many choices not to reproduce would still be seriously immoral. More important, though, are the theoretical difficulties attendant on any claim that prospective (potential or possible) persons have a right to be brought into existence.

Prospective persons are either actual future persons or counterfactual persons. If prospective persons have a right to be actualized (even a prima facie right that would be consistent with Hare's argument), we could only violate that right by violating the rights of counterfactual persons. The children I would have had if I had had children do not exist – they are fictional entities just as Peter Pan and Emma Woodhouse are. We cannot wrong fictional entities because nothing we do can affect them; their pains and sorrows, as well as their joys and pleasures, are as fictional as their existence ([19], p. 265). If I do not have a child, it is not the case that there is a child whom I have failed to conceive and give birth to – there is no such child. Since there is no such child, I could not have violated any right it might have had to come into existence, nor could I have wrongfully ignored its interests – as a nonexistent being, a counterfactual person no more has interests than it has rights. One might conclude from this that, if we take an action with the result that no next generation exists at all, then we have not done a wrong because there is no next generation to have wronged. This need not follow. I will argue later that we can establish a serious obligation to produce a next generation without recourse to rights or other moral claims of prospective persons.

Future persons are quite another matter. We can wrong them because things that we do can affect them ([6], p. 148; [19], p. 271). Suppose that I construct a time bomb to go off one hundred years from now and place it in the middle of Washington, D.C.; in the year 2084, it explodes, killing a thousand people. I killed those people, even though I am no longer alive at the time that they die. Although I did not intend to have killed those particular people – I have no way of knowing who they will be – the same will be true of someone who sets a bomb to go off in the same place tomorrow. Admittedly, my culpability will not make much difference at the time of the deaths – I will not be around to be tried and put in jail. However, it does make a difference to my decision now. I would consider myself as much a murderer for placing a bomb with a 100-year fuse as one with a one-day fuse, and just as obligated not to do it. The only difference might be the (possibly misguided) hope that, given the extra time, someone would figure out how to defuse it (or possibly that there would not be any people left to kill by that time).

There is one more objection left to be answered here. Suppose we are dealing with a decision where conception has already taken place, but birth has not. A pregnant woman discovers through prenatal tests that the fetus has a genetic defect. Is not the fetus an actual, present person? Again, the answer is no, but at this point it becomes more complicated. A fetus or embryo[3] is merely a potential person – that is, something that now exists that may or may not become a person, depending on decisions yet to be made. Note that I am here (as I have been throughout) using the term "person" in a technical sense, that is, as a being that possesses whatever characteristics are necessary for having rights or being due moral consideration. I would argue that a minimum necessary condition for being due moral consideration is the capacity for some mental life – something without any mental life cannot have pleasures and pains, desires or needs ([19], pp. 267–270). Amniocentesis is performed late enough in pregnancy that the fetus may well fit this minimum condition; this would require us to take into account the pain an abortion might inflict on the fetus ([17], pp. 118–119), but newer techniques such as chorion biopsy, which are performed earlier in pregnancy, may eliminate or reduce this problem. Moreover, this minimum condition is not sufficient to establish a right to life – painlessly killing a being that is conscious and capable of feeling pain but is not capable of having desires about the future does not frustrate any desires or plans which that being is now capable of experiencing ([17], p. 78, 102). Michael Tooley has argued that the right to life or the wrongness of killing is conceptually connected to the capacity to have a desire to continue to live (he calls this capacity "self-consciousness") [18]. If we restrict the term "person" to those beings with self-consciousness (and, thus, a right to life, if one is talking in terms of rights), then a fetus is, from the time of conception, potentially a person, but will not be actually a person until some time after birth.

It may be tempting to say that, since the fetus is potentially a person, it has a right to life (the Potentiality Principle). But that which exists now is not a person; it does not have the characteristics of a person, it is not self-conscious and does not have desires that killing it would violate.

Moreover, the principle that we ought to grant rights to those who only potentially have those rights is clearly untenable when applied to rights in general. That a two-year-old child is potentially a rational adult who can and should make her own decisions not only does not make it morally necessary to grant her the right to liberty now; it does not even make it morally excusable. If granted the right before she has the relevant capacity, she may seriously harm herself. The relevant capacity for the right to liberty is a capacity to exercise the right as well as the capacity to desire the object of the right. An incapacity to desire the object of the right or to desire things dependent on it (as in the fetal incapacity to have desires about the future) is, on the other hand, an incapacity to be harmed in the relevant respect. Whichever relationship the criterion for possession of a right has to the right, if the criterion is *the* condition (or a necessary condition) for possession of the right, then a merely potential possessor of that condition is, by definition, not a possessor of the right. To say that a mere potential for x is a sufficient condition for possession of a right to y entails that x is not the criterion for possessing a right to y, but that something else which is shared by x's and potential x's is the appropriate criterion. Any claim of this sort would require a separate argument. Note that my argument does not entail the claim that a potential to have a right is morally irrelevant (see also [6], p. 147). Children, who have the potential to become the kind of rational agents that possess a right to liberty, may have rights related to that potential (for example, the right to education); what they cannot have, as long as their capacity is merely potential, is the right to liberty itself. Moreover, I am inclined to think that in this case (and in other cases of putative rights related to potentiality) the obligation is not an obligation to the present child but to the adult she will become (education is something that children often do not desire and, in the case of very young children, may be incapable of desiring).

The apparent radical moral difference between merely prospective persons and potential persons is illusory. If the fetus is only a potential person and not an actual person, decisions about abortion will be morally similar to decisions about birth control. There will be some differences: a fetus in the later stages of pregnancy is probably capable of feeling pain and is due the moral consideration required for all sentient beings; we know more about potential persons than we do about merely possible persons, and the chance of our being able to turn them into actual persons is greater. Moreover, if we do carry the pregnancy to term and raise the child, there will be a person who may be harmed or benefitted by what we do to the potential person, but if the fetus is aborted, that person will never exist (all that existed was something that might have been a person) and, as I argued above, counterfactual persons do not have rights.

This argument, of course, depends on the acceptance of what has now become the standard philosophical analysis of the problem of abortion. Those who reject such arguments on theological or other grounds would argue that we do at least have obligations to presently existing human fetuses. But such a rejection presupposes an alternative account of the morally relevant criteria for the possession of the right to life, which would extend the right to life to human fetuses. Recent investigations (e.g., [6], [17], [18], [19]) suggest that this task will be difficult, if not impossible, unless we are willing to grant rights as well to nonhuman animals (and, if we include zygotes, to one-celled animals). However, extending the right to life in this way makes the ecological problems of continued population growth even more morally serious; if we grant most animals rights to life, then the growth of the human population seriously threatens the rights of more than the human population.

RIGHTS OF FUTURE PERSONS

Unlike counterfactual persons, future persons can be harmed. However, claims about reproductive decisions harming future persons have a paradoxical nature. Suppose, for example, we are talking about a couple, Leon and Vivian, who are going in for genetic counselling because their families have a history of a genetic disease – Tay-Sachs disease. Their intention is to prevent the birth of a baby with Tay-Sachs and they think of their concern as a concern for the baby as well as for themselves. That is, they think it would not be right for them to bring someone into the world with such a painful and hopeless defect. If they had a child with this disease, they would feel bad for the child and they intend to prevent the suffering of this disease by preventing the birth of a diseased baby. They undergo prenatal tests and abort those fetuses with Tay-Sachs until they conceive a child who does not have the defect.

The problem that arises is this: it is clear that Leon and Vivian's use of genetic technology has prevented suffering – if they had had the Tay-Sachs babies, those babies would have died painfully with very little in the way of positive experiences to counterbalance the pain. But have they avoided wronging those babies or violating their rights not to suffer? Suppose they had not aborted, once they knew that Vivian was carrying a fetus with Tay-Sachs. The baby is born; they name him David. They knowingly brought a defective baby into the world, knowing that he would live a short life and die painfully. Have they wronged David or violated his rights? The claim that by bringing David into existence or by allowing to happen a disease that could only have been prevented by his nonexistence we could be violating David's rights or harming him may seem paradoxical. If a person's identity is determined by his genetic makeup, then it would appear that any correction of a genetic defect that changes the person's genetic nature would be a change of the person. Hardy Jones classifies genetic defects as nondetachable and argues that giving a child a genetic endowment of a particular sort can never be a violation of that child's right because it is not possible to respect that alleged right: if the child is not born, he does not exist to have a right, and if the genetic makeup of the fetus is altered, we have a different child ([8], p. 249). Presumably this objection would not extend to a phenotypic cure of the defect – a solution that does not alter the genotype of the individual, but masks the action of the genetic defect to create a nondefective phenotype.

There are two distinct problems here. One arises if we use prenatal diagnosis and abortion, eliminating individuals with defective genotypes before they are born. A somewhat different problem would arise if we had the technology for genetic surgery, such that we could manipulate the genes at conception so that the resulting individual is genetically identical to the original individual with the exception of the defective gene (the degree of genetic and phenotypic alteration this would require would depend on the genetics of the disease). I will argue that a genetic disease may be a wrong to the individual who has it if (1) it could have been prevented by either of the above technologies and (2) someone deliberately and knowingly refrained from so preventing it.

In the story related above, I think it is legitimate to claim that David has been wronged. Knowingly producing unnecessary suffering is wrong. David is suffering. His suffering was produced by an action (or series of actions and omissions) which the people who performed them knew would produce that suffering. One might be tempted to claim that the wrongness of the suffering is overridden by the fact that the only alternative to that suffering would be something even worse for David, that is, his nonexistence. Even if my argument that the nonproduction of a prospective person is never a wrong to that person fails, this argument assumes that the life that David is living is worth the pain he suffers. In the case of Tay-Sachs disease, I find this claim implausible. However, there are many genetic defects where this is by no means as obvious – for example, the lives of mildly Down's syndrome individuals may not be at all a misfortune to them – in such cases, the denial of wrong may depend on the impossibility of wronging counterfactual persons.

However, suppose we switch our focus from the individual to the next generation as a whole. There is a limit to the number of individuals we can produce without reducing the quality of life for those individuals. The existence of genetically defective individuals reduces that number even farther, since such individuals place a disproportionate strain on the earth's resources ([8], p. 253). Imagine a situation in which the quantity of the next generation is set at 100 individuals. Imagine further that, if we do not use the genetic technology that we have, 15 of them will have genetic defects ranging in seriousness from the mild – e.g, hemophilia – to the quickly lethal such as Tay-Sachs, and that by the use of prenatal tests and abortion we can reduce that number to 3 out of 100. Setting aside the possibility that we have also unknowingly eliminated some benefits, it seems clear that the generation produced by the second policy is better off than the generation produced by the first one. There is less suffering and no one has been wronged (since we cannot wrong counterfactual persons). I think that the generation produced by the first policy would have a legitimate complaint against their ancestors, knowing that their ancestors could have prevented some of their suffering by following the second policy. Even if we reject this conclusion on the grounds that the people who are suffering would not have existed if their ancestors had followed the second policy and therefore cannot very well complain, it is clear that the ancestors who followed the first policy brought into being more suffering (without a corresponding increase in happiness) than did those who followed the second policy. Moreover, if it would be peculiar for the defective generation to complain about the suffering they could not have existed without, it would not be similarly peculiar for the nondefective generation to be grateful to their ancestors for their use of the genetic procedures that produced their felicitous genetic endowment. Thus, the ancestors who follow the second policy are producing more moral good even if the ancestors in the first instance are not violating anyone's rights. The example exaggerates both the number of defects (which would be closer to three by current estimates) and our capacity to detect them (which is currently closer to half), but the moral point is the same.

I have argued that, even if the only alternative is prenatal detection and abortion, the knowing production of a genetically defective individual is a prima facie wrong to that individual. I think the case is even clearer if the alternative is genetic surgery. For it is easy to imagine an individual in such a case being grateful that his parents had had genetic surgery performed. Such a person would not be concerned about not being the same person his parents conceived, nor should he be. It is not at all clear that genetic surgery would be any more likely to create a different person than would a treatment that masks the destructive effects of the gene. To go back to our case of David with TSD, imagine two possible (presently nonexistent) technologies: in scenario 1, genetic surgery is performed to eliminate the genetic defect; in scenario 2, an effective means of replacing the enzyme missing in TSD is used. Either scenario would produce a David who is radically different from the Tay-Sachs infant. They would be much more radically different from the Tay-Sachs infant than they would be from each other on any criterion of personal identity other than the arbitrary one of identity of genetic makeup. A failure to use either of these technologies would be equally a wrong to David.

If my analysis is right, then the apparent paradox in the claim that bringing a genetically defective individual into the world wrongs that individual is only apparent. If we take future generations into account in reproductive decisions, then one of our goals will have to be to prevent such wrongs. I will argue next that we cannot avoid such wrongs by failing to reproduce altogether.

THE CLAIM OF SOCIETY: OUR OBLIGATIONS ABOUT AND INTEREST IN FUTURE GENERATIONS

We normally think that reproduction is an important social function. We say that our children are our future, and assume that the elimination of the species or of the society would be a catastrophe beyond measure. Because it appears to contradict these values, the claim that by deciding not to produce persons we can render them nonexistent and therefore immune to any wrongs we might otherwise do to them appears suspect. Part of the problem may be one of language; it is important to keep in mind that there is no "them" that we eliminate in rendering "them" nonexistent – "they" never existed in the first place. If we only owe obligations to actual persons and can prevent there being actual persons, then how can we have an obligation to create a next generation? The next generation, if we (effectively) decide not to produce it, will not exist, therefore not have rights, and therefore not be wronged by our decision not to produce it.

I think that the solution to this problem is to say that the wrong that is done by a failure to produce a next generation is a wrong to us rather than a wrong to them (who do not exist). If the next generation is our future, then an event that prevents their existence robs us of our future. We need not indulge in metaphysical contortions of discussing the wrongs done to counterfactual persons by preventing their existence in order to account for the catastrophic nature of the end of the species. If the species or the culture is engaging in enterprises that are inherently valuable, then its elimination eliminates those values (possibly permanently) ([2], pp. 66–67; [9], pp. 196–197). Moreover, the elimination of either the species or a culture would cut short enterprises that actual (present and past) persons have and have had a stake in. Annette Baier argues that we are, as moral beings, necessarily members of a cross-generational community and that, as such, "we have no right to decree the ending of an enterprise to which we are latecomers" ([1], p. 178). People in the past built cities and universities, governments and societies, arts and farms and families in the expectation that they would be carried on by and benefit people in the future. If we end the human enterprise at this point, their expectations and hopes will have been futile and we will have failed in an obligation to them.

Indeed, the assumption that the elimination of a future generation must be a wrong to them rather than to us is only conceivable on an extreme atomist view of human beings and of human interests. Strict utilitarians and other atomists may find themselves in the position of arguing that the extinction of the species may only be wrong in that those people who are around during the inevitably slow process will have a significantly less happy life than they would have had otherwise because they have to bear the burdens of being the last generation (for example, there will be no one around to help them when they grow old) ([13], p. 44). I think it is legitimate to find this analysis bizarre because it significantly misrepresents the human enterprise and human desires. Even when most of our lives are taken up with immediate concerns that do not go beyond ourselves or our lifetimes, typically those desires and goals that people regard as giving their lives value, as making life worth living, are goals that in some way or other go beyond one's own lifetime. Just as looking back on the history of our culture or family lends significance to a life, looking to the future gives meaning to life. Eliminating a next generation will automatically frustrate this whole category of significant human desires, from artists' desires to produce art that lasts, to parents' desires that their children have a better life than they do.

For these reasons, all of us have an interest and a stake in there being a next generation. One might also argue that, as beneficiaries of the ongoing process of the generations, we have obligations to contribute to its continuation. However, I think it would be a mistake to conclude here that an individual decision not to reproduce is morally wrong or even not, morally speaking, private. It would only be in very unusual circumstances that a decision not to reproduce would be, or would significantly contribute to, a decision to end the human enterprise. Were such circumstances to occur, the obligation to contribute to the continuation of the species might override the moral concern for the privacy of the decision. In some instances a decision which, as an individual decision, appears private because the effects on others are trivial will, if repeated in sufficient number, have quite serious consequences. Ordinarily, a decision to make one's contribution to the enterprise in nonreproductive ways would be unproblematic – indeed a concern for overpopulation might suggest that the good of the enterprise depends on a significant number of people either not reproducing or limiting their reproduction. However, when the survival of the species – or, more probably, the survival of a valuable genotype – begins to depend on individual decisions, the society acquires a significant interest in those decisions. It is improbable that in the near future (absent a catastrophic event such as nuclear war) such a consequence will hang on any small number of individual decisions. However, concerns of this sort might legitimately prompt social policy to encourage and make easier decisions that would contribute to the preservation of valuable genotypes. For example, concern for lower reproductive rates among highly educated women (on the assumption that education is correlated with intelligence, which is to some degree genetic) might prompt support of measures such as paid maternity leave and child care in jobs requiring high levels of education in order to encourage reproduction (or at least reduce the social and economic penalties for reproduction) on the part of those women.

Not only does the society have an interest in the existence of the next generation, but it also has an interest in the genetic quality of the next generation. Whether or not we could establish a social interest in increasing valuable genotypes (such as those related to intelligence), it is clear that certain genetic defects (those that make individuals dependent on social resources while being noncontributors to those resources) are, beyond our concern for the well-being of those who have them, of significant interest in that producing them will place a substantial burden on the resources of the society.

RIGHTS OF PROSPECTIVE PARENTS

I have been arguing that both decisions to reproduce and decisions not to reproduce may be morally wrong. This would appear to suggest that, if there is a right to reproduce, it is not without moral restrictions. None of the arguments above discusses the parents' stake in reproductive decisions, yet I started this investigation with the assertion that reproductive decisions should be treated as private decisions of the parents, because the stake that the parents have in the decision is significantly greater than that of anyone else. The parents' interests fall into three categories: (1) the stake of the pregnant woman in her body; (2) the stake of both parents in the need to care for the child once born; and (3) the stake of both parents in the continuation of the family. The first of these considerations is relevant to reproductive decisions to the extent that they affect pregnancy and birth; the second will be relevant to the extent that the society assigns the care of children to biological parents; the third category will be of concern as long as people's hopes and expectations are tied up with the continuation of a family where "family" is genetically defined.

The fundamental limiting factor on social control of reproduction lies in the fact that decisions about reproduction are typically decisions about what will go on in and what use will be made of a person's body. Since every reproductive decision affecting pregnancy is a decision about women's bodies, granting them rights in making such decisions is necessary if we are to take women seriously as moral agents. However, to grant someone the right to make a decision is not to suggest that she can never make a morally wrong decision. As I have argued above, some reproductive decisions are morally wrong and are – unlike most medical decisions about oneself – wrong because of their effect on the interests of others. Why, then, should we grant women the right to make these decisions, if they may adversely affect others' interests?

First of all, a negative reproductive decision is a substantially private decision, except in the extreme case where the future of the species or genotype is seriously threatened. In the extreme cases, but not in the ordinary ones, the stake the society has in the decision might justify overriding individual rights in order to prevent it. Thus, with this exception, such a decision cannot be seriously morally wrong. Moreover, to prevent negative reproductive decisions is to force positive ones, that is, to force the woman to undergo pregnancy against her will. I have argued elsewhere [10] that to do so is to treat her body as the object of other people's rights, thus to treat it as an object, and, thus to treat her as less than a person. Even on the assumption that the fetus is a person (and that abortion, therefore, is not a private decision), to prohibit abortion on the grounds of the fetus's right to life would be to treat the woman's body (and thus the woman) as the object of the fetus's rights rather than as the subject of her own rights.

However, the argument that a woman has a right to continue a pregnancy is weaker than the argument that she has a right not to continue it. Reproducing creates a person who can be harmed by being brought into existence with a sufficiently serious defect. Independent of any right the prospective parents might have to reproduce, there are two main objections to forcibly preventing prospective parents from producing children with genetic defects. First, any coercive method (such as compulsory sterilization or abortion) requires the involuntary invasion of a person's body. This leaves us with noncoercive methods – policies that encourage and facilitate genetic family planning – and with infanticide. Even on the assumption that the newborn is not a person (as many philosophers writing on abortion (e.g., [17], [18]) have felt themselves forced to conclude), infanticide is an unattractive option. The other problem is a more openly political one. The question of what (if any) genetic defects are sufficiently serious that their possession would support a claim of wrongful birth is a difficult one and one on which it would be futile to expect sufficient agreement for legislation in the near future. I have offered arguments for the claim that there could be such a thing as a morally wrongful birth, and I find these arguments morally persuasive – I think that, if I were Vivian in my example above and deliberately continued a Tay-Sachs pregnancy, I would think of myself as having done something morally wrong. However, as strong as I regard the moral force of the arguments, they would need to be substantially more morally weighty to support coercive measures overruling prospective parents' own values. Even if we did, at some future date, have as much agreement about wrongful births as we do now about the desirability of, for example, compulsory vaccinations, I might note that even forcing life-sustaining medical treatment for children whose parents voice religious objection has proved controversial for our courts. Thus, although decisions *to* reproduce are not strictly speaking private, there are good reasons for treating them as if they were private decisions of prospective parents, even apart from the fact that the person whose stake in the decision makes them not private does not exist at the time of the decision (even though the prospective person could not take part in the decision, his or her interests could be represented as those of children are) ([6], p. 142).

The second consideration of parents' interests (the need to care for the child once born) applies as long as the task and the expense of such care falls on them. The care and raising of a child is a substantial part of a parent's life project and makes demands on both income and activity. The care of a child with genetic defects may raise those demands dramatically and may interfere with parents' ability to fulfill not only their own needs but also their other responsibilities ([4], p. 43). This concern might be removed from the parents' interests by state supported child care, medical care, and even institutionalization for such children, but the latter measure ignores the child's need for individual affection, and all of these measures merely shift the burden – and, therefore, the legitimate interests – from the parents to other members of the society. Once there is a person who needs such care, a social policy that spreads the burden more equally and reduces the demands on individual families is (if it preserves the child's interests) desirable because it is more equitable. However, such a policy (and the moral responsibility that exists independent of that policy) creates a legitimate social interest in whether or not people will be produced with these needs. Such an interest might suggest a justification for legislation against such births, but these considerations would be outweighed by the problems with coercion raised above. I might note here that, in the absence of a policy of social support, this second consideration (but not the first consideration) would give parents a legitimate interest in decisions about infanticide or about withholding medical treatment from defective newborns. On the assumption that newborns are persons, this would not be an overriding interest, but if newborns are not fully persons, it might be.

The final consideration that demonstrates parental interest in reproductive decisions (the stake that parents have in continuing a family) is a variety of the kind of interest I have argued we all have in the next generation. To the extent that our long-range goals are predicated on the existence of a future generation, we have an interest in the bringing-into-existence of that generation. Similarly, to the extent that an individual's goals are predicated on the existence of a future generation with a related genetic makeup, that individual has an interest in the creation of genetically related persons. We may also have obligations to preceding generations to continue the family. However, we can hand down traditions, values, habits and ways of life to nonrelated (for example, adopted) children. Even to the extent that the continuation of the family is conceived of as requiring biological continuity, genetic family planning might be consistent with it, but only if selective reproductive decisions can be made. A family with a history of hemophilia might be happy with an eradication of the genes that are related to that disease, but it is only by selective methods that such an eradication can be consistent with genetic family continuity. At present, a genetic defect for which there is only genetic screening but no prenatal test poses the choice of either giving up genetic family continuity (through preventing reproduction altogether) or taking a chance on reproducing the defect.

THE INTERESTS OF PHYSICIANS AND HEALTH PROFESSIONALS

It should be clear by now that, on my account of the relevant interests, the power of physicians to affect reproductive decisions is out of proportion to their interest in and right to have a say in such decisions. This is not, by any means, a moral criticism of physicians; it is rather a comment about technology and the way it is organized. Technology gives us control over reproduction (as it gives us control over production), but that control may not be in the hands of those who need it. Just as the control of energy lies in the hands of the owners of the sources of energy and in the hands of those who understand the technological processes of producing energy, rather than in the hands of the consumers of energy, so the technology of reproduction rests in the hands of hospital administrators and physicians rather than in the hands of the reproducers. Some feminists have argued that the more reproduction (and, in particular, the process of pregnancy) has been technologized, the less control women have had over it, first, because the expertise and the social control of health care delivery moved from female midwives to a predominantly male medical profession ([5], Ch. 3), and, second, because the more technologized pregnancy becomes, the less active the pregnant woman becomes in the process of delivery – the more she becomes an object to be operated on ([15], Ch. VII). The movement toward technocracy is more morally dangerous in the case of reproduction than in the case of production because, in production, the workers and the consumers are persons, but not the products or the means of production, while, in reproduction, the products (children) and the means of reproduction (women) are also persons.[4] Thus their technological control of reproduction places physicians in a more morally delicate position than that of the engineer who has control over production, even without taking into account the fact that the physician has more control over reproductive technology than engineers do over productive technology [11]. Licensing laws give physicians monopolistic control over the delivery of health care and the use of medical technology ([3], p. 97), and physicians may have substantial say over hospital policy. Licensing laws have been used to prevent midwives and nurses from practicing independently and hospital boards have organized to prevent the delivery of reproductive technology (abortion and related techniques) within their area of purported delivery of service by preventing access to hospital use to physicians who perform abortions. Thus physicians have the power to prevent substantial numbers of people from gaining access to the medical technology necessary to make reproductive decisions effective. And they use this power. Perhaps some of this power could be democratised or returned to the patient through a different organization of health care and of health care delivery, but what remains for us now is the fact that, given the present structure of health care, physicians will be in a position to make or influence choices about other people's reproductive lives.

One might argue that, since the physician's power to make these decisions comes in part from a politically granted monopoly and in part also from socially maintained institutions (medical schools and hospitals), to that extent the physician's power is a trust (like a politician's power) and, in order to be legitimate, must be exercised for the benefit of the constituents of that power [11]. This principle may not be sufficient where there is a conflict of values rather than of interests; to the degree that the political power of physicians is monopolistic we cannot simply discharge the physician with discrepant values and replace him with someone whose values are consistent with our own.

Does the physician have any moral claim on reproductive decisions? Although the substantial effects of reproductive decisions will fall on others (the parents and the possible child), the practice of medicine is the physician's life activity and to grant the decision to others would appear to give them control over the physician's life, and might require individual physicians to perform actions contrary to their sincere principles. However, few people have control over their own life activities. Few jobs allow any latitude in deciding how one's talents and abilities are to be used and those that do restrict it by also requiring attention to the needs and interests of the people being served. The physician who is required to follow his patient's wishes is not worse off than the architect who is required to follow his client's wishes or the factory worker who is required to follow the directions of management.

The problem appears to lie in the nature of the moral conflict. When physicians take as their goal the preservation of life – regardless of whether or not that is in the patient's best interest or consistent with the patient's wishes – they are claiming a higher goal than that of serving their immediate clients. If we take such physicians' objection to the use of reproductive technology as a serious moral position (rather than as a form of sexist or religious prejudice), we are left with a seemingly insoluble problem – or at best a choice between two evils. On the one hand, we might force people to perform actions to which they have serious moral objections. On the other hand, if we do not do this, we may be allowing professionals to prevent others from gaining access to something they have a right to as patients, and, moreover, by our cooperation, we all become responsible for and morally culpable for the harms caused by their refusal to prevent suffering that is technologically preventable.

I argued above against using coercive laws to prevent potential parents from producing children with serious genetic defects. I would use the same arguments against physicians making decisions for the parents. The fact that the coercion comes from the medical profession rather than the government may make it more, rather than less, objectionable – we have a chance to vote on legislators. However, an individual physician's refusal to perform an abortion or other procedure may not be coercive; whether or not it is depends on the circumstances and the availability of alternatives. Even a hospital's refusal may not be coercive (it may be one hospital in a large city with many hospitals of equal cost and availability performing the procedure); it is the monopoly that creates the coercion.

CONCLUSION

For any decision there are two kinds of moral issues: one has to do with questions about who should make the decision and the other with what decision ought to be made (and, thus, indirectly, with who has to be considered in the decision). Resolving the question of who has a right to make the decision does not solve questions about what the decision should be – a person who has a right to make a decision may make a morally wrong decision. Ordinarily, having a morally legitimate interest in the decision gives one some moral claim to a say in the decision. Reproductive decisions raise two kinds of problems for this general schema. First, those who have the power (through control of reproductive technology) to make certain kinds of reproductive decisions – that is, physicians, who are capable of either preventing or enabling people to carry out the technological procedures – do not have a major stake in those decisions. Second, one category of persons that will be crucially affected by reproductive decisions is not only not available for consultation, their existence as persons is contingent on the decision itself.

I have argued that (although one can imagine circumstances in which it would be otherwise) the interests of the medical profession, the society, and the prospective children are not sufficient to override the considerations in favor of protecting reproductive decisions as private decisions to be made by the patient – that is, by the present person whose body the decision is about. Since most (but not all) reproductive decisions relate to pregnancy, this will have the consequence that the final say for most reproductive decisions should rest with women. However, in order for a particular decision to be a moral one, it must take into account the interests of all who would be affected. Thus, the morally responsible woman will take into account in making the decision the possibility of harm to a person she might create, the interests of the prospective father, the possible burden on or contribution to society of the prospective person, and the interests of the relevant medical professionals, as well as her own needs and responsibilities.

Rollins College,
Winter Park, Florida, U.S.A.

NOTES

[1] It should be noted here that Mill's definition of the private sphere is that it has to do with those actions directly affecting only the person acting and other adults only with their consent ([12], p. 11). However, in most medical decisions, the person acting is the physician. Thus, strictly speaking, medical decisions are private only in cases where there is no disagreement between the patient and physician. However, if we regard the physician as the patient's agent, and therefore the physician's actions as the patient's actions, medical decisions will be private. I will put this problem aside until the section on the interests of physicians.

[2] "Prospective person" is both more general and more specific than the usual term, "potential person". A possible person (here I use "possible person" as the most general term) is a prospective person relative to a particular decision or set of decisions. "Potential person" suggests that there is now some thing that will or may become a person, and I will reserve it for that use. My great-great-grandchildren are possible persons who may be prospective persons with respect to some decisions, but they are not potential persons (in this restricted sense) and may never be potential persons.

[3] There is a tradition in philosophical ethics (perhaps an objectionable one) of using the term "fetus" generically, without distinguishing between "fetus" and "embryo". In cases dealing with prenatal detection, this practice is less likely to be misleading than in more general discussions of abortion, for although new techniques such as chorion biopsy may push the stage at which we can detect abnormalities in utero back into the first trimester of pregnancy, they are still performed later than eight weeks and, thus, we are dealing with the abortion of a fetus rather than an embryo. The arguments that follow (about fetuses) will apply *a fortiori* to embryos and zygotes.

[4] I have argued this point in a comment on 'Marxism, Feminism, and Reproduction' by Susan Rae Peterson, at a meeting of the Society for the Philosophical Study of Sex and Love held in conjunction with the Eastern Division Meetings of the American Philosophical Association in Boston, December, 1980.

BIBLIOGRAPHY

1. Baier, A.: 1981, 'The Rights of Past and Future Persons', in [14], pp. 171–183.
2. Bennett, J.: 1978, 'On Maximizing Happiness', in [16], pp. 61–73.
3. Bledstein, B.: 1976, *The Culture of Professionalism*, W. W. Norton and Co., New York.

4. Bolton, M.: 1979, 'Responsible Women and Abortion Decisions', in O. O'Neill and W. Ruddick (eds.), *Having Children: Philosophical and Legal Reflections on Parenthood*, Oxford University Press, New York, pp. 40–51.
5. Ehrenreich, B. and English, D.: 1978, *For Her Own Good: 150 Years of the Experts Advice to Women*, Anchor Press, Garden City.
6. Feinberg, J.: 1981, 'The Rights of Animals and Unborn Generations', in [14], pp. 139–150.
7. Hare, R. M.: 1975, 'Abortion and the Golden Rule', *Philosophy and Public Affairs* **4**, 201–222.
8. Jones, H.: 1981, 'Genetic Endowment and Obligations to Future Generations', in [14], pp. 247–259.
9. Kavka, G.: 1978, 'The Futurity Problem', in [16], pp. 186–203.
10. Ketchum, S.: 1984, 'The Moral Status of the Bodies of Persons', *Social Theory and Practice* **10**, 25–38.
11. Ketchum, S. and Pierce, C.: 1981, 'Rights and Responsibilities', *The Journal of Medicine and Philosophy* **6**, 271–279.
12. Mill, J. S.: 1978, *On Liberty*, Hackett Publishing Co., Indianapolis.
13. Narveson, J.: 1978, 'Future People and Us', in [16], pp. 38–60.
14. Partridge, E. (ed.): 1981, *Responsibilities to Future Generations*, Prometheus Books, Buffalo.
15. Rich, A.: 1977, *Of Woman Born: Motherhood as Experience and Institution*, Anchor Press, Garden City.
16. Sikora, R. and Barry, B. (eds.): 1978, *Obligations to Future Generations*, Temple University Press, Philadelphia.
17. Singer, P.: 1979, *Practical Ethics*, Cambridge University Press, Cambridge.
18. Tooley, M.: 1973, 'A Defense of Abortion and Infanticide', in J. Feinberg (ed.), *The Problem of Abortion*, Wadsworth, Belmont, pp. 51–91.
19. Warren, M.: 1981, 'Do Potential Persons Have Rights?', in [14], pp. 261–273.

LISA SOWLE CAHILL

ON THE CONNECTION OF SEX TO REPRODUCTION

THE "NATURE" OF HUMAN SEXUALITY

What is the "nature" of human sexuality? Is it or is it not essentially connected to the conception of offspring? And, if it is, how precisely is the meaning of that modifier "essentially" to be specified? When we ask about the "connection" between sex and reproduction, are we inquiring about a physiological fact or a moral norm? Or, to pose the problem another way, what is the connection between the "fact" that sex can be reproductive, and evaluative norms specifying what human beings "ought" to pursue in their sexual and parental lives?

In the Western philosophical tradition, heavily influenced by and even intertwined with the Judaeo-Christian religious tradition, presuppositions about the connection of sex to reproduction, or to "procreation" (to use a term more at home in the humanities), have been defended or reexamined in the light of certain conceptions about "the nature of things", especially the natures of humans and human society.[1] No process of reflection on human nature, of course, proceeds without a social and cultural context; some concrete, historical setting is its very condition of possibility. At the same time, such a process has little *raison d'être* if it is not also supposed that the human mind can perceive some essential human characteristics not sheerly relative to a particular historical and social epoch or place. This is not to say that "human nature" is some ahistorical and objectively self-evident reality, impervious to the changes and contingencies of actual experience. However, it is to presuppose that, despite its pliability and variability, human experience is not without continuity. It is in an effort to unburden as far as possible reflection on human sexual nature from the accretions of centuries of relatively uncritical assumptions about what sex is "for" that some philosophers have turned recently to more "phenomenological" examinations of its nature.

Earl E. Shelp (ed.), Sexuality and Medicine, Vol. II, 39–50.

A common focus is the dimensions of sex that are oriented primarily to the communion and fulfillment of the partners, rather than those dimensions oriented equally or more to the welfare of the species (reproduction), or of society (marriage and family). It is as a language of interpersonal expression that the meaning of sex is elaborated most frequently.[2] What is striking is the degree to which the authors of many such analyses try to circumvent traditional biases about the meaning of sex by narrowing their vision almost entirely to the immediacy of sexual experience, especially of sexual intercourse. One of the more acute examples of such a method is A. Goldman's argument that purposes such as reproduction or love are "extrinsic" to the nature of sex; sexual activity is the pure and simple attempt to satisfy "the desire for contact with another person's body and for the pleasure which such contact produces" ([10], p. 268). A question that presents itself is whether such a focus also represents a certain culturally sustained ethos tending both toward individualism and toward empiricism. The norm of the good is the individual; the criterion of the true is empirical validation, particularly through sense experience.

Even so, Goldman and others represent positively the perception that if we are to understand the nature of human sexuality, the primary phenomenon to which we must attend is the experience of being sexual. Other authors tend to portray sex as at an essential level the expression of some personal relationship [13], [16], [17], [25], [27]; or at the least, as a relation of mutuality in fulfilling one another's sexual desires [4]. In these interpretations, sex as physical enjoyment receives significant attention, but usually is seen as attendant upon an affective interpersonal relationship, even if not a permanent commitment. It has been observed often enough that sex pursued for the *sole* purpose of physical enjoyment and satiation is likely to become technical, depersonalized, and ultimately boring ([10], p. 1587).

As appreciation of sex's nature as intimacy increases, support for defining its nature as also reproductive declines. Why is this the case? An obvious explanation is that intimacy can occur appropriately in contexts inappropriate (or even incapacitated) for reproduction. It is also true that in humans the desire for coitus occurs outside of the proportionately small period of female fertility. A less obvious but perhaps more profound reason may lie in the very method of assessing the significance of sex. Generally speaking, conception or "reproduction" is *not part of the immediate experience of sexual relations*. The *focus* of attention in sexual acts, especially intercourse, is much more likely to be one's own sexual experience and that of one's partner, and the physical and affective relation of the couple, rather than the likelihood of conception. Exceptions to this rule are those relatively infrequent instances in the *ensemble* of the average person's sexual acts when he or she desires that conception occur, or, conversely, is acutely fearful of conception. The grounding of definitions of the meaning of sex in phenomenologies of the sex act tends to result in the view that sex essentially is "for" intimacy alone, since conception rarely is part of the phenomenon under analysis, namely, the subjective conscious experience of the sexual partners during sexual acts, especially sexual intercourse. The analysis concludes to a rejection of the traditional or conventional view that reproduction also is part of the intrinsic meaning of sex (i.e., at least a part of what sex is "for" essentially). It also leads to a repudiation of the received moral norm congruent with the latter definition, which prescribes that "sex for intimacy" is only or most appropriate when its context is appropriate at least minimally for conception (i.e., the partners are able to be personally and together accountable for the welfare of offspring, should they result).

It hardly needs to be pointed out that this exclusive focus on sex as intimacy has not prevailed universally. There have been (and continue to be) eras and societies in which each individual and each couple wants for social and economic reasons to conceive, bring to birth, and raise, as many children as possible. There also have been (and continue to be) cultural and religious traditions in which sexual faculties are perceived as existing naturally for the primary or even sole purpose of reproduction. The intention of procreation is then taken as necessary to the legitimate satisfaction of sexual drives, and to the legitimate sexual expression of marital love. In fact, not only the human sexual capacity, but even the female herself, as the "vessel" of male seed and "mother", may be seen by both sexes as existing primarily to serve the end of procreation (cf. [9], [19], [24]).

I want to suggest that analyses limited for the most part to the immediate personal experience of sex do highlight its most distinctively human dimension, the consciously interpersonal one, but do not account for its total essential reality. In this, sex parallels other realms of human experience, e.g., other forms of expression and symbolization, such as language and art. To understand the human significance of the phenomenon as a whole we must examine its relevance to the human community as well as to the individual, then inquire in what ways these two dimensions of meaning mesh or are differentiated. Persons are not just individuals, but members of a species, and procreation of their kind is one mode of their connection to the dignity, purposes, and welfare of the species, and to the ecology of the universe.[3] Of course reproductive sex in the twentieth century will hardly serve the requirements of universal ecology by sheer numerical increase. Quite the contrary. However, sexuality remains a crucial basis of the ties of blood relation and conjugal kinship undergirding the structures of societies everywhere, whether in the form of clan, tribe or family, "nuclear" or "extended."

Much confusion about sex and reproduction (especially its moral aspects) originates in two reductionist tendencies. On the one hand, the typical subjective perception of immediate sexual experience is liable to be substituted for the total reality of that experience; on the other, the assumption frequently is made that what is "normative" for the sexuality of the whole species ("human" sex) also is normative for all sexual acts and relations of individuals ("individual" sex or sex "of the couple"). The former assumption leads often to the conclusion that since, in individual experience, the affective or even merely physically pleasurable experience is foremost, the indispensable norm of individual sexual acts is affection (which may or may not be seen as accompanied normatively by some degree of commitment), or even merely reciprocal pleasure. *Therefore*, it is concluded, reproduction cannot be a normative meaning of sex, but something to be transcended, to be "liberated from," and so forth. In the introduction to their collection of essays on the philosophy of sex, R. Baker and F. Elliston take a position of this sort. They propose that since the purpose of sex is erotic fulfillment, which does not require permanence, the "erotic act itself could be unburdened of the onus of possible parenthood" ([2], p. 5).

On the other hand, the assumption that the sexuality of the species sets the norm for that of individuals tends to the conclusion that since human sex is affectively unifying and produces offspring, *therefore* all individual sex acts should be simultaneously unitive and procreative.[4] A key and difficult question regards the extent to which, if any, the meaning of sex for the species should govern, that is, constitute a moral norm for, the sex lives of individuals. Even if certain purposes do in fact characterize sex if taken as an activity of the species, is there therefore any obligation on the part of the individual to seek to realize these meanings personally? In other words, can a personal moral "ought" be derived in any sense from what globally "is" the case? While philosophers and theologians alike have learned to be wary of uncritical derivations of moral norms from statements of fact, there is still widespread (and, I think, justified) support not only for the idea that the ideal for which persons ought to strive is in some way congruent with essential human nature, but also for the idea that what is "essential" and "normative" in human experience is fundamentally related to what human societies generally are inclined to cherish, institutionalize, and promote. Without such presuppositions, both traditional and current interest in discovering the real "nature" of sex would be senseless. I think it not implausible to suggest that there ought to be a *prima facie* presumption in favor of the essential characteristics of human sex as normative for individual sex. If there is, though, this presumption certainly can be overridden by factors in individual situations that count against the successful integration of procreation into sexual relationships and commitments. Why? Because personhood is more distinctively human than biology (reproduction). In humans, freedom and rationality control and direct biology and the physical conditions of life in community. This is what is meant by saying that humans are moral agents at all.

Even so, in twentieth century Western (especially North American) culture, the interpretation of the interpersonal dimension of sex as an agreement of two (or more) free, autonomous, and self-gratifying individuals has become pervasive to a remarkable degree (cf. [2], [4], [17], [26], [27]). This may be due partly to the predominance in our cultural ethos of the liberalism rooted in the social contract theories of Locke, Hobbes, and Rousseau, and embodied in the American constitutional tradition of civil liberties. In the liberal view, individuals are seen as free agents who are not intrinsically interdependent, and related through innate social and moral bonds, but who enter into "consenting adult" relationships wherein they agree to duties of noninterference or of reciprocal support in the pursuit of their own ends. (This point is made very effectively by D. Callahan in an essay entitled, appropriately, "Minimalist Ethics" [6].) Such a view of persons and their relations, including sexual ones, is of course at odds with the way most of the human race for most of its history has perceived the human sexual capacity and the social implications of its tendency toward procreation. The normative question is whether the liberal ethos represents progress or regress in respect to human understandings of what human community is about essentially.

I want to offer a counterview, premised like many other contemporary understandings of sexuality, on a "phenomenological" appreciation of it. While my assessment of what is to be "seen" about sex and reproduction is, no doubt, no more "value free" than that of the liberal philosophers with whom I have disagreed, I hope to draw attention to at least one element in the human sexual "situation" that has not been appreciated fully. That element is *corporeality* (cf. [18]). It will not escape the attention of anyone who thinks or writes about sex that humans are not only affective, rational, and volitional creatures, but also embodied ones. Corporeality or embodiment is a precondition of the sexual relationship as interpersonal, and an essential dimension of sexual experience and sexual acts themselves. S. Ruddick effectively ties together the corporeal and affective dimensions of sex in her definition of "complete sex" as "mutually embodied, mutually active, responsive desire" ([25], p. 97). Complete sex acts, in Ruddick's view, are morally superior, conducive to love and respect, and order desire into a social relation that solves the fundamental moral tension between narcissism and altruism.

The association of sexual morality with mutual pleasure and intimacy is valid, but I want to suggest that to recognize the embodied nature of sex will take us beyond this association. The sexual capacity of the human body is just as obviously conducive to reproduction as it is to reciprocal sexual pleasure. The meaning of human sexual function as embodied is not exhausted by affective and emotional and physical satisfaction, though those are essential constituents of its meaning, and perhaps even the most important ones. The fullness of the human sexual reality is neglected if we simply overlook in normative analysis the fact that the creation of offspring is one "natural" purpose of sex as a physical phenomenon, and that the consolidation of a relationship (the conjugal and parental one), and of a social group conducive to the nurturance and education of offspring (the family) are among the purposes for which the affective and interpersonal dimensions of sexuality are apt. The intergenerational family (which can and does assume many forms in many cultures) is a physical and interpersonal structure of connection between the sexuality of the individual and the community. As others have noted, it is communal interdependence, in both its physical or material and interpersonal aspects, which is critically neglected in "liberal" accounts of human nature (M. Bayles makes a similar point in [5]).

The social or communal meaning of sex probably has been its primary one in most periods of human history. If the individual and interpersonal meanings of sex now gain ascendancy rightfully, this does not require that the biological meanings of sex be severed from the essential and normative human sexual reality. As the philosopher M. Midgley has insisted on the basis of her survey of ethnographical studies, it is no denigration of the human to observe that it is also "animal." She reminds us that "Our dignity arises within nature, not against it" ([15], p. 196), and presses us to answer, "Why should a narrow morality necessarily be the right one? Why should not our excellence involve our whole nature?" ([15], p. 204).

P. Ricoeur offers an analysis of sex in which its procreative meaning is retained but transformed by the modern "alliance of spiritual and carnal in the person" ([23], p. 135). The primitive imagination, according to Ricoeur, endows sexuality with meanings symbolic of the divinities immanent in inanimate nature; ethical monotheism defines the sacred as transcendent, and reinterprets sexuality as only a part of the total order, a part having no meaning apart from its social contribution of procreation. In the contemporary ethics of sex, the union of both immanent and transcendent aspects of the sacred in the person must be respected: "by the control of procreation, reproduction ceases to be a destiny, while at the same time there is a liberation of the dimension of tenderness in which the new idea of the sacred is expressed" ([23], p. 135). The "ethic of tenderness" takes over the institution of marriage, transforming its primary goal of procreation so that procreation is included in sexuality rather than vice versa, and so that "the primary aim of marriage is the interpersonal relationship" (p. 137).[5]

MEDICINE AND THE CONNECTION OF SEX TO REPRODUCTION

In a sense, there is no question of any role of the ethos of the medical profession or of medical techniques in "establishing" or "severing" some optional connection between sex and reproduction. This connection is a given in the full human experience of sexuality or of human sexual "nature". The fact that human nature must be akin to that of other animal species, and must likewise be more than that to which humans can consent freely, should be appreciable by practitioners of the medical arts in particular. The key question is the way medical professionals perceive, present, and control reproduction in relation to other purposes and outcomes of sex. Medicine (here I mean physiological rather than psychological medicine) of its nature deals more with the physiological than the affective or properly interpersonal aspects of sex. The physician, for example, might contribute to the health of sexual function and response, thereby enhancing the physical base of personal enjoyment and interpersonal expression. He or she might also focus on fertility, either limiting or improving the potential of coitus for it, perhaps indirectly serving the inclusive society of couple or family.

In undertaking to realize these possibilities, the medical practitioner should be aware both that such interventions do have moral implications and that there is no such thing as a "value-free" approach to sexual function. What, then, should the attitudes of those in medicine be toward the normative status of the fact that sex "naturally" has certain outcomes, among them reproduction? Should all sexual acts be open to this end? Should the end of reproduction by technical means ever be separated in any particular sex act or relation from the end of interpersonal union? When and why and for whom? These questions are moral ones, and involve philosophical or theological evaluation; they will not be settled by any amount or kind of empirical physiological or anthropological or sociological evidence taken alone.

Unfortunately, it is impossible to isolate any single universally persuasive answer to these moral questions. While some philosophical and religious traditions regard both human sex and individual sex as intrinsically reproductive, concluding that sexual relations should only occur in contexts where reproduction would be appropriate, others have offered that even if reproduction characterizes the meaning of the sexuality of the species, it does not yield a specific norm for every individual act of sexual intercourse ([21], [7], Ch. 2). However, given the fundamental character of the links both social and physical between sex and reproduction, it seems unrealistic, individualist and even dualist, to draw a facile conclusion that the connection of sex to reproduction is incidental; that reproduction is something superadded or appended to the essential reality; and that the practice of medical specialties related to sexuality might proceed as though conception, pregnancy, and birth were occasional consequences which neither practitioner nor patient need contemplate further than the provision of relatively effective contraception.

CONTRACEPTIVE AND REPRODUCTIVE TECHNOLOGIES

This will be a brief and illustrative application of theory to concrete practice. Traditionally, medical practitioners have had to face the problem of contraception (and of abortion[6]) as a means of limiting fertility. Moral warrants plausibly offered in favor of contraception are that the elimination of the consequence of reproduction can enhance opportunities to seek intimacy through the sex act, and, more generally, that additional births can jeopardize the welfare of couples, families, and communities. New on the medical and moral scene are techniques such as artificial insemination by husband or donor, in vitro fertilization, and birth by "surrogate" womb, in which circumvention of the sex act itself, alone or in combination with circumvention of intimacy of relation between biological parents, can enhance opportunities for reproduction.

In most theories of sexual ethics, sex, love (affective bonding and commitment), and reproduction are treated as three variables ideally accompanying one another. The new "twist" is that it is now reproduction that is made to take place without sex, rather than sex that takes place without conception. Justifying reasons most frequently offered are the desire, need, and even right of individuals or couples to bear children when they are otherwise biologically incapable of doing so; and the interest of individuals, couples, and societies in promoting the births of genetically "healthy" or even "superior" babies. Moral legitimation of these methods depends at least on the premise that no moral wrong is committed in the separation, at least temporarily, of sex and reproduction.

The question posed by donor methods of medically-assisted conception is whether love too legitimately can be separated from the reproductive constellation. Donor methods present us with a disjunction between conception and the responsibility of "parenthood" in the fuller long-term sense. Interestingly, while both AIH and AID are common in the U.S., and IVF (usually between spouses) is becoming more so, there is significantly greater reluctance at both the popular and theoretical levels to accept methods of artificial conception that do not involve the biological cooperation of two partners committed not only to each other, but also to the rearing of the child that results, than there is to accept the employment of the same techniques by spouses. R. McCormick, a moderate Catholic theologian, speaks of donor techniques as an "intrusion" into "the marital covenant and family" ([14], p. 1459), and concludes that,

> Furthermore, the majority opinion of ethicists has been that AID, by the introduction of donor semen, separates procreation from marriage, or the procreative sphere from the sphere of marital love, in a way that is either violative of the marriage covenant or likely to be destructive of it and the family ([14], p. 1458).[7]

The importance of love as a context both for sexual acts and for procreation is visible through the much greater acceptance of reproductive techniques involving a man and a woman who in the totality of their relationship (if not in the event of conception itself) unite love, sexual acts, and reproduction. It is the interpersonal aspects of sex, then, which, as Ricoeur points out, are primary in contemporary sexual ethics. This primacy also is evident in the respect and value accorded the sexually expressed love of couples who do not have children. The moral conclusion that follows is that it is easier to argue that the needs and desires of individuals and couples are weighty enough to override the natural and generally desirable conjunction of the three variables when it is sex that is dissociated from reproduction in a context of committed love (AIH, nondonor IVF), than when coital reproduction takes place outside of that context (unwed parents), or when both love and sex are separated from reproductive cooperation (AID, surrogate womb, donor IVF).

The consequences for the connection between sex and reproduction are twofold. First, beyond the conjunction of sex and reproduction as human meanings of sexuality, there is a *presumption* in particular cases in favor of their union in at least the totality of a committed, loving, sexual relationship, a presumption grounded in the natural bodily integration in human sexual intercourse of pleasure, intimacy, and conception. But, second, since intimacy is a more distinctively human meaning of sex than procreation, the latter is a secondary good in relation to the former and may for sufficient reason be eliminated from a particular sex act, or even from the relationship as a whole.

Boston College,
Chestnut Hill,
Massachusetts, U.S.A.

NOTES

[1] In Judaism and Christianity, proposals about human nature rest ultimately on antecedent affirmations about God as Creator and, in some accounts, as Redeemer. God is said to have created humanity with certain characteristics and purposes, to have restored or instituted certain human possibilities through his savings acts, and to command or require certain responses from humans, which it is their "nature" as created and redeemed to give. For an excellent critical history of sexual ethics in the West, with particular attention to religious traditions, see [9].

[2] For religious examples, see [11] and [12]; for philosophical ones, Kant in [13] and [3], especially the section, 'Sex and Morality'.

[3] D. P. Verene stresses that the transcendent and communal meaning of sex is *eroticism* (rather than reproduction) ([28], p. 112).

[4] This has been the traditional Judaeo-Christian position, though even the unitive aspect has been submerged until relatively recently (cf. [10]. An example of a recent statement of this position is the 1968 encyclical of Paul VI, *Humanae Vitae* [20].

[5] For an eloquent theological defense of the intrinsic character of the sexual act as at once love-strengthening and life-giving, see [21].

[6] Abortion is an exceedingly important moral problem, but I do not intend to discuss it here, for its moral implications certainly extend beyond the permissibility or impermissibility of limiting fertility.

[7] In the years since McCormick made this statement, "majority opinion" has begun to shift, at least among those closely involved with the practice of fertility therapy. Reservations about donor methods have been set aside in favor of individual autonomy and of effective pursuit of a medical solution to infertility. An example is a 1986 report of the Ethics Committee of the American Fertility Society. A key value is "procreative liberty." The use of donor sperm, donor eggs, and donor preembryos is accepted, although surrogate motherhood is not ([8.] The conclusions are summarized on pp. 765–785.) As a member of the Committee, McCormick was allowed to register his objections in an Appendix ([8.], p. 825).

BIBLIOGRAPHY

1. Arthur, J. (ed.): 1981, *Morality and Moral Controversies*, Prentice-Hall, Englewood Cliffs.
2. Baker, R. and Elliston, F.: 1975, 'Introduction', in [3], pp. 1–28.
3. Baker, R. and Elliston, F. (eds.): 1975, *Philosophy and Sex*, Prometheus Books, New York.
4. Baumrin, B. H.: 1975, 'Sexual Immorality Delineated', in [3], pp. 116–128.
5. Bayles, M. D.: 1975, 'Marriage, Love, and Procreation', in [3], pp. 190–206.
6. Callahan, D.: 1981, 'Minimalist Ethics – On the Pacification of Morality', *The Hastings Center Report* **11**, No. 5, 19–21.
7. Curran, C. E.: 1979, *Transition and Tradition in Moral Theology*, University of Notre Dame Press, Notre Dame.
8. The Ethics Committee of the American Fertility Society, *Ethical Considerations of*

the New Reproductive Technologies, published as Supplement 1 of *Fertility and Sterility: Official Journal of the American Fertility Society* (September 1986, Vol. 46, No. 3).
9. Farley, M.: 1978, 'Sexual Ethics', *Encyclopedia of Bioethics*, Vol. 4, Macmillan, pp. 1575–1589.
10. Goldman, A. H.: 1977, 'Plain Sex', *Philosophy and Public Affairs* **6**, 267–287.
11. Guindon, A.: 1976, *The Sexual Language: An Essay in Moral Theology*, University of Ottawa Press, Ottawa.
12. Jeanniere, A.: 1967, *The Anthropology of Sex*, Harper and Row.
13. Kant, I.: 1963, *Lectures in Ethics*, L. Infield, trans., Harper and Row, New York.
14. McCormick, R. A.: 1978, 'Reproductive Technologies: Ethical Issues', *Encyclopedia of Bioethics*, Vol. 4, Macmillan, New York, pp. 1454–1464.
15. Midgley, M.: 1978, *Beast and Man: The Roots of Human Nature*, Cornell University Press, Ithaca.
16. Moulton, J.: 1975, 'Sex and Reference', in [3], pp. 34–44.
17. Nagel, T.: 1975, 'Sexual Perversion', in [3], pp. 247–260.
18. Nelson, J. B.: 1978, *Embodiment: An Approach to Sexuality and Christian Theology*, Augsburg, Minneapolis.
19. Ortner, S. B. and Whitehead, H. (eds.): 1981, *Sexual Meanings: The Cultural Construction of Gender and Sexuality*, Cambridge University Press, Cambridge.
20. Paul VI, Pope: 1968, *Humanae Vitae*, United States Catholic Conference, Washington (English version).
21. Ramsey, P.: 1975, *One Flesh: A Christian View of Sex Within, Outside, and Before Marriage*, Grove Books, Nottinghamshire.
22. Reich, W. T. (ed.): 1978, *Encyclopedia of Bioethics*, 4 vols., Macmillan, New York.
23. Ricoeur, P.: 1964, 'Wonder, Eroticism, and Enigma', *Cross Currents* **14**, 133–141.
24. Rosaldo, M. Z. and Lamphere, L. (eds.): 1974, *Woman, Culture, and Society*, Stanford University Press, Stanford.
25. Ruddick, S.: 1975, 'Better Sex', in R. Baker and F. Elliston (eds.), *Philosophy and Sex*, Prometheus Books, Buffalo, pp. 83–104.
26. Slote, M.: 1975, 'Inapplicable Concepts and Sexual Perversion', in [3], pp. 261–267.
27. Solomon, R.: 1975, 'Sex and Perversion', in [3], pp. 268–287.
28. Verene, D. P.: 1975, 'Sexual Love and Moral Experience', in [3], pp. 105–115.

H. TRISTRAM ENGELHARDT, JR.

HAVING SEX AND MAKING LOVE: THE SEARCH FOR MORALITY IN EROS

IS SEX SPECIAL?

Sexuality engenders moral problems because it serves radically different kinds of goals. Reproductive sexuality focuses on bearing children and involves moral concerns about their healthy birth, as well as their successful physical, emotional, and social development. Societies have different, well-developed social institutions for reproductive sexuality and its consequences, usually ensconced within special social roles such as those of being a parent, a guardian, or a child. Sexuality also plays an important social role. Many have argued that the character of homo sapiens was influenced by the tendency of members of the species to form fairly stable and enduring male-female dyads marked by numerous non-reproductive acts of intercourse [5; 10; 12; 16]. Unlike many other species of primates, large harems are not the rule and complex social relations usually develop between one male and one female human. Sex makes a significant contribution to this pair bonding [22]. Finally, sexuality plays a recreational role. This is attested by the age-old presence of bordellos, catamites, and gigolos. Human males and females have sought sexual recreation apart from their commitments either to reproductive sex or to enduring social bonds.

It is not just that one can distinguish among the three clusters of goals supported by human sexuality. The goals are often in conflict. The optimal mate, from the point of view of raising healthy children well adapted to their environment, may not be the same individual who is available as one's optimal social mate. Adoption, artificial insemination by donors, and *in vitro* fertilization with the use of surrogate mothers are all examples of practices through which individuals outside of the central sexual social bond are crucially involved in achieving the reproductive goals of sexuality. Though these different clusters of goals can

Earl E. Shelp (ed.), Sexuality and Medicine, Vol. II, 51–66.

be orchestrated in harmonious fashions, such is not always the case. Biological and adoptive parents can come into conflict, even to law suits. The recent literature in bioethics contains reports and speculations concerning the possibly complex tensions and conflicts that may be engendered by such practices. Consider the difficulties that may arise when a wife has her ova fertilized by donated sperm so that the zygote produced can be transferred to the uterus of a surrogate mother who promises to return the child to be raised by that wife and her husband [14, 28]. One need not turn to modern technology for conflicts between the social and recreational elements of human sexuality. Gossip at any bar or bridge game or a few moments spent listening to soap operas will provide sufficient examples of men and women seeking satisfaction from recreational sex outside of their ongoing commitment to a particular sex partner.[1]

In all of these conflicts one might very well ask what is morally unique. Surely there are crucial moral issues at stake with regard to truth-telling, promise-keeping, and the fashioning of contracts. But in principle, one might think, sexuality offers no unique moral problems – just more interesting examples. Speculations about the circumstances under which John may be excused from spending the promised weekend with Mary in San Antonio tend to engage more prurient interest than exploring the practice of promise-keeping with respect to when John is excused from returning Mary's copy of Palestrina's Pope Marcellus Mass. In both cases the analyses will need to focus on the moral significance of promise-keeping and on exception-making circumstances, Does a prior promise to hunt white-tailed deer near Boerne take precedence over John's promise to Mary? Does he owe some compensation to Mary for not having more carefully arranged his commitments? Perhaps he should at least offer a week together on Padre Island to compensate for the loss of conviviality and consortium in San Antonio. May Mary demand to be taken along on the hunt, even if John's hunting partners wanted a weekend for themselves, wildlife, and beer *sans* consorts? If John goes hunting, may Mary go with George to San Antonio, despite Mary and John's mutual commitment not to have other sexual partners, on the ground that John's violation of his promises defeats John's claims against her? What if, instead of George, she threatens to go with Sally for a lesbian weekend in Fort Worth? The analyses of such webs of responsibilities regarding sexual matters, both inside and outside of marriage, would appear to involve no unique moral issues. The moral character of monogamous, polygamous, polyandrous, and even more complex sexual alliances, heterosexual and homosexual alike, would seem to depend on the character of the commitments made and the likely benefits and harms to ensue, without any claims having been advanced about moral issues unique to sexuality. The only exception may be that of reproductive sexuality, which involves the production of a third unconsenting party.

Still, one must acknowledge a substantial moral literature that argues a contrary position. The Judeo-Christian tradition in general, and Roman Catholicism in particular, allege moral problems unique to reproductive, social, and recreational sexuality, contending that certain forms of sexual activity are intrinsically unnatural and therefore immoral.[2] Other activities (e.g., adultery), though not unnatural, have been decried for violating divine law.[3] Such considerations have led to the condemnation of certain activities focused on the achievement of reproductive sexual goals. Here one must include the condemnation of artificial insemination by a donor (A.I.D.), which has been forbidden by the Roman Catholic Church because it involves (1) the procurement of semen by masturbation, an unnatural, sinful act, as well as (2) adultery, the mating of individuals not joined in valid marriage ([18], pp. 228–244; [19], pp. 90–96).

Analyses that conclude that certain activities are intrinsically unnatural presuppose (1) that one can discern in nature designs and (2) that in addition it is immoral to reshape or violate those designs. Within an understanding of nature that was more teleological than contemporary science, this view of design and its moral significance were more credible. However, the more one sees nature's laws and structures arising from blind, surd processes of physical and biological evolution, the less plausible it is that one can discern designs. One may grant that it is likely that the human penis developed under selective pressure favoring a structure that efficiently supports ejaculation in vaginas, while still holding that its development may have also been guided by other selective pressures. It is probable that hearing developed to allow animals to avoid predators, to discover prey, to find mates, and to locate and respond to offspring in need. In addition, it is possible that, as mammals in general and primates in particular engaged in play and recreation, the survival of groups was enhanced by the solidarity created through enjoying forest sounds together or engaging in non-reproductive sex. In fact, Theodosius G. Dobzhansky and others have argued that the social character of homo sapiens developed in part because freedom from estrus allowed human sexuality to take on an importantly social if not recreational valence [8].

Such interpretations of the development of the human marriage bond and family made by the loss of estrus are contested by a number of individuals. As Donald Symons indicates in a review of the literature, the evidence is sufficiently ambiguous to allow more than one reading ([27], pp. 97–141]). He proposes two. The first scenario suggests that estrus was lost because of the advantage to females of continuous advertisement of sexual availability. Once the loss of estrus was sufficiently complete, males could no longer trade the spoils of the hunt for immediately reproductively successful sexual intercourse. A new strategy of courting a particular female in order to gain her long-range reproductive loyalty may then have developed in order to maximize the male's reproductive success. Males would then come to look for indices of attractiveness other than immediate reproductive capacity. This scenario might be termed the bride price scenario for the development of human sexuality.

The second scenario, unlike the first, depends on estrus having been lost after marriage bonds had been fairly well established. Under such circumstances, the subsequent loss of estrus would have allowed females to bond with the best individual available as a permanent partner, thus maximizing her social advantage (e.g., availability of food and protection), while allowing the same female to mate with a different male who had superior biological advantages for her offspring, even though he was already socially mated with another female. This scenario might be called the infidelity model ([27], pp. 138–141).

As Symons stresses, neither of these two scenarios, nor any that has occurred to any writer so far, may actually provide an accurate account of how estrus was lost and human inclinations for bonding developed. It is extraordinarily difficult, if not impossible, to reconstruct the evolutionary past in matters such as these. At least it seems safe to conclude that non-reproductive sexual activity has played a significant role in the development of primates in general and humans in particular. As a result, it is impossible in a biological sense to identify sexual reproductive acts as *the* natural sexual acts, while excluding non-reproductive acts, which may have been favored because they contributed to human fitness, and are thus from an evolutionary perspective just as natural.

For this reason among others, when one considers how different elements of our bodies evolved, it is difficult to identify *the* function.[4] Our ears, fingers, tongues, and sexual organs can be regarded as having numerous functions. They and our dispositions to use them are likely to have developed under very complex selective pressures. To term sexual organs simply "genitals" is to prescind from this adaptive range of significance, which includes their also being sociables and pleasureables. Yet within the context of medieval Christendom's neo-Aristotelian framework for understanding design in nature, it became plausible to isolate a particular function as *the* orthodox or canonical one. The difficulty with such interpretations of human sexuality is that they depend on a teleological view of design in nature that is not defensible through empirical science. Moreover, the premises required for concluding to moral prohibitions are not available through general secular reason alone. Special religious or metaphysical assumptions are required to show that it is immoral to tamper with or contravene an established design in nature, if such in fact exists. One could even grant that God designed the structure of the human body and still question, absent further premises, whether the Deity meant the given design to be morally normative. One might contend that the intent of the Deity was to provide a starting point for man's office as co-creator so that the manipulation of human nature is a laudatory act that fulfills human destiny. In short, much more is required to identify certain sexual acts as unnatural and therefore immoral – perhaps even a commitment to a particular religion.

One is confronted with a problem integral to secular value theory: it is difficult to find general concrete canons for moral guidance. As a consequence, the questions that are clearly decideable through general secular moral reasoning tend to involve procedural issues concerning the respect due to persons. This is because the choice of any particular, concrete value hierarchy presupposes a particular moral sense, the choice of which in turn requires a higher level moral sense, *ad indefinitum*. In contrast, respect of persons focuses on a minimal condition for the possibility of morality ([11], chap. 2 and 3). The difficulty in determining the canonical concrete view of the good life has similarities with the problem of determining what should count as diseases or perversions, though the latter problem can also be put in somewhat different terms. After Darwin, it is no longer possible unambiguously to specify good adaptation in terms of a standard environment. There is no such environment, as there was with Eden, to which one can make reference independently of particular sociocultural and environmental conditions. In that evolution has no goals in the sense of moral purposes, but rather selects those traits that maximize fitness, humans are left with the problem of deciding which goals they wish their bodies and minds to subserve. As a result, persons must choose both the environment and the goals in terms of which adaptation and maladaptation, health and disease, are to be understood ([11], Chap. 5). These considerations apply as well to human sexuality, insofar as one wishes to consider particular sexual preferences diseased or perverted. Unless goals can be discovered in nature itself or by appeal to concrete generally defensible value standards, the goals for sexuality must be provided by individuals themselves or by particular moral and aesthetic traditions. The foregoing considerations give grounds for abandoning the hope that avenues are open for discovering generally justifiable concrete goals for sexuality.

THE SEARCH FOR MORAL STANDARDS FOR SEXUAL BEHAVIOR

These sorts of reflections support the removal of homosexuality as a prima facie mental disorder from the *Diagnostic and Statistical Manual of the American Psychiatric Association* and its replacement with ego-dystonic homosexuality. That is to say, a homosexual has a mental disorder, according to the 1980 nosology of the American Psychiatric Association, only if changing sexual orientation is a matter of persistent concern to that person ([3], p. 281f). This represents a development from the first edition, which classified homosexuality under the general rubric, "sociopathic personality disturbance," as a form of sexual deviance, and the second edition, which considered homosexuality under the general rubric of "non-psychotic mental disorders" ([1], p. 38f; [2], p. 44). The result is a subjective standard that does not make an appeal to concepts of perversion or natural sex.

The problem is that once one is bereft of arguments from natural law, and as long as there are not generally defensible secular moral canons for sexual behavior, it becomes increasingly difficult to pick out as diseased, disordered, or immoral sexual activities that do not involve unconsented-to force against the innocent. Sexual mores seem to go aground on the crisis of nihilism because it is difficult morally to rule out sexual activities that many, if not most, would hold to be deviant, perverted, or at least impoverished (e.g., necrophilia or bestiality[5]). This general controversy has been joined by Robert Spitzer, who chaired the Taskforce on Nomenclature and Statistics of the American Psychiatric Association, which created *DSM-III* and its category, ego-dystonic homosexuality. Why, for example, should an individual with a shoe fetish not be accorded the same liberty as a homosexual who can reject the diagnosis of a psycho-sexual disorder, if the sexual preference is ego-syntonic [26]? After all, if one has a fulfilling sexual life with shoes as one's erotic object, one avoids not only the risk of divorce and lovers' spats, but sexually transmitted diseases as well.

Does this mean that all standards are gone? That all forms of sexuality are equally acceptable? Many people find this conclusion uncongenial and have attempted to develop arguments, philosophical and otherwise, to show that certain sexual preferences are better than others [13, 15, 20, 23, 24, 25]. At least some sexual activities can be ruled out on moral grounds insofar as they involved the use of unconsented force against the innocent (e.g., rape) and thus reject the very notion of mutual respect. Some have also attempted to exclude sexual experiences as degraded or immoral, even in the presence of consent, if they involve extremes of domination such as found in sadism ([24], p. 96). Here, however, particular views of proper human relationships are at stake. If the partners are consenting, such relationships do not violate the very fabric of morality. Moreover, there is the disturbing fact that much of our sexual impulses may in fact be tied deeply in our brains to impulses involving dominance and submission ([27], pp. 146–165). Consider the variants of the phrase (and its material equivalents), "screw you," which is often used as an aggressive, hostile invective. However, the same phrase can occur as a part of statements said in an inviting, erotic fashion, such as, "Don't you think it's time to screw?"

Erotic and aggressive impulses are tightly bound, one to the other. This does not mean that such impulses should not be orchestrated in ways other than those to which past evolutionary forces may have inclined us. The ingredient impulses that we possess are sufficiently complex so that submission and domination may not only be seen by some as edifying ("submit to the will of God"), but sexually exciting ("You swept me off my feet," "Take me, I'm yours," "Make me yours," "He conquered me," "That hurts so good," "It is so good when you claw my back when you are excited," "When you tie me to the bed,

I get so excited"). At one extreme of such cases, most may get squeamish, but if the partners find that dominance and submission deepen their mutual commitment and heighten their sexual pleasure, generally defensible retorts are not easy. After all, if one contemplates the character of sexual congress, however pleasureable it might be, it conveys no conceptual necessity but only a brute facticity. One might consider here a variant of the old saying, "Inter excrementum lotiumque nascamur." Does all as a result become simply a matter of taste?

Some have attempted to establish that certain sexual preferences or activities are incomplete in not allowing a full realization of the range of values open through interpersonal relationships [23, 25]. Such considerations lead us to distinguish between what is a successful sexual experience qua experience from what is a successful relation with another person. It may very well be that for some or for many masturbation, shoe fetishism, or first encounters with strangers provide the most exciting and engaging sexual experiences. Still, such success in sexual sensations must be distinguished from the qualities that specifically mark sexual relations with other persons, which are defined at least in part in terms of the quality of the relation. Three sorts of issues are at stake: (1) the qualities of sexual sensations, (2) the qualities of relationships with other persons, and (3) the qualities of sexual relationships with other persons. The third category requires the difficult balancing of the goals of the first and second. One may have a good relationship with mediocre sex or a mediocre relationship with great sex. These considerations are then further complicated by the issues raised by reproductive sexuality. The phenomenology of a fulfilling sexual experience in an extended sense of experience is thus complex.

We are returned to the opening theme regarding the tensions engendered by the diverse goals realizable through sex. Since the person who may make the best available social partner providing the best sexual sensations may not be the best reproductive sexual partner, the recipes for disaster are multiple. Hence, the age-old wives' and husbands' advice to the young: Don't think that the person with whom you want most to rut is necessarily the person with whom you should want to spend the rest of your life, much less have as a parent of your children. The multiple goals of sexuality are difficult to realize within one sexual relationship. The old adages about getting one's priorities straight in such matters testify to the difficulty of harmonizing these goals. A good partner for a one-night stand is not necessarily a good friend, and even a good friend is not necessarily a good parent for one's children.

THE WILL TO VALUES

As Nietzsche has argued, nihilism does not destroy the world of values. It may preclude the discovery of a ready formed constellation of values. However, individuals can still fashion meaning, structure a world. The superman (and superwoman – the German is *Uebermensch*) is able to look at the abyss and laugh because the superman still has the courage to create values in the absence of enduring standards. Friends create a common history, an enduring bond, a meaning to their sexuality. Their sexuality finds its "higher truth," to use a Hegelian phrase, within the embrace of their commitments. They create a world of the mind, an objective place for their spirits, which place sustains, interprets, and gives moral and other significance to the pleasures and urges of sexuality. The stronger the commitment, the greater the need to mediate the sheer erotic within the light of mutual understandings and the conceptual structure of shared ideas. In this way, friends create a world for themselves that transcends the subjective urges of each individual. Commitments objectify and place the spontaneous within a context. Though one may lose the quick life of passion drawn from encounters with novelty and discovery, one gains a meaning that transcends the moment and sustains a sense of enduring significance. In this sense, friendship, including sexual friendship, is an accomplishment, a work of art.

It is also a sacrifice. Since humans are not gods or goddesses, friendship entails the abandonment of other possibilities. Humans have limited resources of energy and time. The person who has a hundred friends has none or is a god. To develop a sexual friendship with one person is not to be able to have a sexual friendship with someone else (I do not mean to exclude polygamy or polyandry, but only to suggest that King Solomon did not know all of his thousand wives as friends). *Omnes determinatio est negatio.*[6] To choose to be the friend of one, two, three, or four individuals is an exclusion of other possibilities and a commitment to making certain common worlds of friendship and not others. Since friendship is a binding of persons, it requires moral and intellectual undertakings. One must find and develop at least some common moral, aesthetic, and intellectual understandings. As a consequence, making love in the extended sense of making a sexual friendship is not just physical but involves moral and intellectual labor.

The matter is even more complex when one assumes the task of making a family. Families transcend the lives and times of any one member. This is the case even with nuclear families. In making a family, friends commit themselves not only to being good lovers and trustworthy partners, but to being good parents. Even nuclear families have their histories and make their traditions. "We always go to Big Bend, all together, every year. The kids camp out in one tent, me and my woman [man] in another. We set time aside to get close to God, ourselves, lust, and Texas." The children as adults then tell the story further. *Making* love, *making* a friendship, *making* a family are tasks of constant commitment and of sacrificing some possibilities on behalf of others.

Extended families are even more complicated. They demand more sacrifice and offer a greater compass of meaning and depth of history. "Ever since the Smiths moved out here to Guadalupe County, they knew the difference between love and lust. They never got no damned divorce if they had kids. If they weren't getting what they needed at home, they might go to the cedar brakes to find it, but they knew what their commitments meant." Extended families have stories to tell. Extended families have characters who serve as exemplars in morality stories. "Uncle Willie Joe left his wife in '08 and went to San Antonio, leaving his wife and kids to fend for themselves. He never came to no good." Or "Aunt Marcia Lou found out too late that her husband was as upstanding as corn after a hurricane. That was her own damn fault. She married him a week after they met at the church dance. But she never did run out on her husband and kids. She might have met with that siding salesman a couple of times in Columbus, but she at least saw that some commitments shoud'nt be broken." Or "Our family has always placed a high value on chastity and commitment. In the past, members of our family did not contract syphilis or get pregnant out of wedlock. In the present and in the future they will not have to worry about AIDS." Whether one agrees with them or disagrees with them, families provide moral frameworks that give and secure meaning and sustain purpose.

Within friendships and families, the desires and urges that characterize sexuality are provided interpretive matrices of ideas and values. The pleasure of skin against skin, the characteristic sexual sensations and pleasures, the relief of sexual tension, the pleasure and discharge of tension that marks orgasm, and the mutual physical affirmation of sexual encounter are given a social place, purpose, and meaning. From masturbation to casual affairs to sexual friendships to the development of families, the mere facticity of sexual sensations and urges is interpreted within ever-richer and more complex webs of values, expectations, and ideas. The seriousness of sexual choices has been underscored by the recent AIDS epidemic, which has caused heterosexuals and homosexuals alike to reconsider casual sexual encounters. But, even after AIDS is preventable and curable in some distant future, *Deus volens*, the challenge of giving significance and meaning to sexuality will remain.

Some may complain that this approach to sexuality supports much of the classic bourgeois sentiments regarding the family and sexual morality. But this fact may provide additional support rather than a criticism. The bourgeois family developed as an institution focused on creating and sustaining resources. Insofar as the aristocracy is free of such needs and the proletariat too impoverished to aspire to them, the bourgeoisie develop and sustain societal resources. Both within and outside of socialist countries such families meet the challenge of sustaining meaning and of husbanding resources, intellectual, moral, and financial. In general, the weakening of the nuclear family through divorce and pregnancy outside of marriage has led to major social strains as more families are headed by a single adult who alone does not command the pyschological, social, and financial resources requisite for the full moral life of a family.

The difficulties involved in sustaining commitments are significant. Sexuality is marked by passions, and passions are not reasonable. They tear at reason, undermine values, and strain the structure of commitments. As Lawrence Durrell had Justine summarize the confusions of passion, "Who invented the human heart, I wonder? Tell me, and then show me the place where he was hanged" ([9], p. 83). One comes to lust after those whom one does not love, and to be driven to love sexually against prior commitments of exclusivity. Because of passion's challenge to constancy and purpose, it is often said that one may hold others accountable for their friends but not for their lovers. It is against such difficulties that love is made in the serious sense of a sexual friendship.

Making love requires creating a place for the passions of sexuality. Commitments are dedications not only of the mind but of the flesh. Friendships habituate our urges. Still, the possible harmonies of mind and body, of intellectual commitment and passion, are multiple. Morality in its general secular mode, as has already been noted, can only give general restraints of truthfulness and faithfulness. It can indicate that promises should be kept, but not which promises should be made. The choice of the character of one's commitments must be drawn from particular understandings of the good life. The morality of sexuality is thus to be understood within two tiers. The first is that of general secular moral concerns regarding autonomy and beneficence. Here there are no uniquely sexual issues, only special passions to bring within general moral constraints. The second tier is that of particular understandings of the good life. This tier includes special understandings of the goods of sexuality. Here issues unique to sexuality may be found. But, since they depend on particular understandings of the good life, they are not reducible to or defensible in terms of general rational considerations. They depend on particular visions of the goods of sexuality, of the ways in which the unique urges and sensations of sexuality are to be orchestrated, experienced, appreciated, and realized against the backdrop of the reproductive, recreational, and social goals of sexuality.

To make love is to make meaning, to fashion a particular history of personal commitment and common action. To have one meaning is not to have another. Enduring relationships among persons are, as all human accomplishments, particular – and they have their particular traditions, however informal. Whatever their weaknesses and justifications, such relationships do not require a deductive proof. They are not necessary truths. They can be avoided. If embraced, they are pursued because of their special claims on us.

A HOMAGE TO APHRODITE: A PRAYER TO ATHENA

Lust can possess us. As the saying goes, when you're hot, you're hot. Even casual observation of human behavior reveals a great deal of rutting in high prurient passion. Aphrodite has her season. But she is not the only goddess. Other powers have their claims as well. Severus Alexander, Augustus and Pontifex Maximus, worshipped in his private chapel Abraham, Christ, Orpheus, and Apollonius of Tyana ([21], XXIX, 2]. The good life requires a balancing of the various divine powers, a successful composition and realization of the human capacities. To underscore this ideal of aesthetic and moral composition, I have interpreted the metaphor of making love to construe love as opposed to lust, as an accomplishment. Sex is for the having. The fabric of love must be created and sustained. Yet the erotic gives direction and allure, it provides pleasures to articulate, it excites the mind and moves the body, it endows flesh and action with the fascination of the seductive – *das ewig Weibliche zieht uns hinan*. Making love requires a balancing dialectic between the demanding urges from Aphrodite and the moral structures given by Athena, the goddess of wisdom. Sex without love is blind, love without the erotic is cold. The fully human requires the grace of both goddesses.

In this essay I have stressed balance and complementarity between the generalities of morality and the particularities of sexuality. This problem presents a general task for the moral life of achieving content within constraints set by reason. Sexuality seems so special, its sensations so unique, and its passions so confounding, that we are invited to believe there are special moral issues associated with the erotic. This prejudice, as has been noted, is supported by the traditional notion that certain sexual activities are unnatural. Aside from special issues raised by reproductive sexuality, involving non-consenting reproduced third parties, unique and generally justifiable moral constraints do not appear to exist. Individuals with different moral senses will have different schedules of benefits and harms. Unless one can establish a particular moral sense as canonical, numerous sexual moralities will exist within the constraints of mutual respect, the necessary condition for the possibility of resolving moral controversies on principle without recourse to force ([11], chap. 2). We are left only with general moral constraints on both the having of sex and the making of love. Those constraints can be discovered in the very character of morality. But such constraints give no content to the moral life. The moral content of sexual life must be created or at least discovered within the context of partially moral traditions or the light of special grace. The general restraints on sexual activity are restraints on persons as such and would apply even to non-human persons, whatever their sexual lives might be. The character of human sexual feeling has a more limited universality and at best is general to primates, mammals, or perhaps higher vertebrates. There may even be special valences of sexual experience tied to the dissimilar roles played by sexuality in the reproductive strategies of males and females, since males maximize their reproductive interest by inseminating as many females as possible, while females maximize theirs by securing support for their offspring. Within general, rationally justifiable constraints, particularity abounds.

Given the vagaries of passion and perhaps even certain innate differences in the dispositions of the sexes, the reconciling of the claims of Aphrodite Pornos and Athena Parthenos in a friendship is a major accomplishment of spirit with flesh. It is like sculpting in the living flesh with chisels of common goals and commitment. Love is made flesh and the inclinations of flesh are rarely to constancy. With sex one rides wild horses. Love is made, not discovered.

Center for Ethics, Medicine, and Public Issues,
Baylor College of Medicine,
Houston, Texas, U.S.A.

NOTES

[1] An indication of the level of successful extra-marital reproductive affairs is given by the results of a study of blood types in a British town, which showed that at least 30% of the children could not have been fathered by the husbands of their mothers ([7], p. 63).

[2] For example, St. Thomas in *Summa Theologica* 2a2ae. 154, 11, argued that sexual acts are immoral when "they are in conflict with the natural pattern of sexuality for the benefit of the species: it is from this point of view that we talk about unnatural vice" ([4], p. 245). In Article 12 he even holds that unnatural acts, all else being equal, are the worst of sins, worse even than rape. "Since, then, unnatural vice flouts nature by transgressing its basic principles of sexuality, it is in this matter the gravest of sins" ([4], p. 247).

[3] Examples are provided by the Biblical condemnation of adultery. Exodus 20:13, Leviticus 20:10, Deuteronomy 15:17.

[4] The proper understanding of functions is debated in the philosophical literature. See, for instance, [6] and [29]. One of the difficulties is that evolution may not favor the character and structure of an organ because of only one "function." Different characteristics of the same organ, as well as the same characteristic's function in different circumstances, may have all favored the development of an organ as it currently exists.

[5] I report here some lines composed by Scholastics inspired by a footnote purportedly added to an early edition of Genicot's *Institutiones theologiae moralis*'s treatment of bestiality: "raro fit cum tigribus." The lines were sung to the tune of St. Thomas's 'Pange Lingua.'

Raro tantum fit cum tigri
Raro cum leonibus
Sed interdum non frequenter
Fit cum elephantibus
Sed vix umquam, imo nunquam
Fit cum scorpionibus.

[6] One might think of Georg Wilhelm Friedrich Hegel's gloss on this phrase taken from Spinoza: "Determination is negation posited as affirmative" ([17], p. 127). Hegel is making the point that to be a determinate anything is not to be something else. To realize oneself as a person is in the end to develop certain possibilities rather than others. To realize love and friendship in one's life is not to experience it simply universally, but rather as love of and friendship with particular individuals. One's loves and friendships have a particularity. They are determined as being what they are in not being other things.

BIBLIOGRAPHY

1. American Psychiatric Association: 1952, *Diagnostic and Statistical Manual: Mental Disorders*, American Psychiatric Association, Washington.
2. American Psychiatric Association: 1968, *Diagnostic and Statistical Manual of Mental Disorders*, 2nd ed., American Psychiatric Association, Washington.
3. American Psychiatric Association: 1980, *Diagnostic and Statistical Manual of Mental Disorders*, 3rd ed., American Psychiatric Association, Washington.
4. Aquinas, St. Thomas: 1968, *Summa Theologiae*, trans. Thomas Gilby, McGraw-Hill, New York.
5. Barash, D. P.: 1977, *Sociobiology and Behavior*, Elsevier North-Holland, New York.
6. Boorse, Christopher: 1976, 'Wright on Functions', *The Philosophical Review* **85**: 70–86.
7. Ciba Foundation Symposium 17: 1973, *Law and Ethics of A.I.D. and Embryo Transfer*, Elsevier, Amsterdam.
8. Dobzhansky, Theodosius: 1962, *Mankind Evolving*, Yale University Press, New Haven, Ct.
9. Durrell, Lawrence: 1968, *The Alexandria Quartet: Justine*, Faber and Faber, London.
10. Eibl-Eibesfeldt, I.: 1975, *Ethology: The Biology of Behavior*, Holt, Rinehart and Winston, New York.
11. Engelhardt, H. Tristram, Jr.: 1986, *The Foundations of Bioethics*, Oxford University Press, New York.
12. Fairbanks, L. A.: 1977, 'Animal and Human Behavior: Guidelines for Generalization Across Species', in M. T. McGuire and L. A. Fairbanks (eds.), *Ethological Psychiatry: Psychopathology in the Context of Evolution Biology*, Grune & Stratton, New York, pp. 87–110.
13. Goldman, Alan: 1977, 'Plain Sex', *Philosophy and Public Affairs* **6**, 267–287.
14. Gorovitz, Samuel: 1985, 'Engineering Human Reproduction: A Challenge to Public Policy', *Journal of Medicine and Philosophy* **10**, 267–274.
15. Gray, Robert: 1978, 'Sex and Sexual Perversion', *The Journal of Philosophy* **75**, 189–199.
16. Hamburg, B. A.: 1978, 'The Biosocial Basis of Sex Differences', in S. L. Washburn and E. R. McCown (eds.), *Human Evolution: Biosocial Perspectives*, Benjamin/Cummings, Menlo Park, pp. 155–213.
17. Hegel, Georg Wilhelm Friedrich: 1965, *Wissenschaft der Logik*, Friedrich Frommann Verlag, Stuttgart-Bad Cannstatt.
18. Kelly, Gerald: 1958, *Medico-Moral Problems*, Catholic Hospital Association, St. Louis.

19. Kenny, John P.: 1962, *Principles of Medical Ethics*, Newman Press, Westminster.
20. Ketchum, Sara Ann: 1980, 'The Good, the Bad and the Perverted: Sexual Paradigms Revisited', in A. Soble (ed.), *The Philosophy of Sex: Contemporary Readings*, Littlefield, Adams and Co., Totowa, pp. 139–157.
21. Lampridius, Aelius: 'Severus Alexander', in *Scriptores Historiae Augustae*.
22. Masters, William and Johnson, Virginia: 1970, *The Pleasure Bond*, Little, Brown and Company, Boston.
23. Nagel, Thomas: 1969, 'Sexual Perversion', *Journal of Philosophy* **66**, 5–17.
24. Ruddick, Sara: 1975, 'Better Sex', in R. Baker and F. Elliston (eds.), *Philosophy and Sex*, Prometheus Books, Buffalo, pp. 83–104.
25. Solomon, Robert: 1974, 'Sexual Paradigms', *Journal of Philosophy* **71**, 336–345.
26. Spitzer, Robert L.: 1986, 'The Diagnostic Status of Homosexuality in *DSM-III*: a Reformulation of the Issues', in H. T. Engelhardt, Jr., and A. Caplan (eds.), *Scientific Controversies: Case Studies in the Resolution and Closure of Disputes in Science and Technology*, Cambridge University Press, New York, pp. 401–15.
27. Symons, Donald: 1979, *The Evolution of Human Sexuality*, Oxford University Press, New York.
28. Walters, LeRoy: 1985, 'Editor's Introduction', *Journal of Medicine and Philosophy* **10**, 209–212.
29. Wright, Larry: 1973, 'Functions', *Philosophical Review* **82**, 139–168.

SECTION II

SOCIETY, SEXUALITY, AND MEDICINE

JOHN DUFFY

SEX, SOCIETY, MEDICINE: AN HISTORICAL COMMENT

The radical change in sex morality that has occurred in the past thirty years is radical only insofar as the past 150 years are considered. Before the 1830's sex was viewed in a much more relaxed manner than was the case even as late as World War II. To most Americans the word "Puritanism" is identified with the colonial period and connotes sexual repression and inhibition. The explanation for this belief can be found in the writings of 19th–century historians who ascribed to Puritans their own attitudes towards sex. The Victorian preoccupation with sex affected all aspects of life, and Victorian historians can scarcely be blamed for reflecting prevailing views.

Whatever the case, beginning in the 1930's a new picture of the New England Puritans began to emerge. The New England settlers were products of Merry England, and although they emphasized godliness, this did not mean that they could not enjoy life. They recognized sex as a major force in life and sought to channel it rather than dam it up. In dealing with sex they were anything but ascetic. John Cotton in 1694 declared in a sermon on marriage that God had little use for those who believe in "the Excellency of Virginity" since he had provided the first man with a wife. He further denounced "Platonic love" as "the dictates of a blind mind. . . ." Another Puritan minister, the Reverend Samuel Willard, several times denounced "that Popish conceit of the Excellency of Virginity" ([31], pp. 591–607).

Church and court records fully bear out the assertions of the ministers. A good many individuals were expelled from their churches and punished for refusing to have sexual relations with their spouses. The major limitations placed on sex were that it must come within the marriage framework, must not come before religion, and must be abstained from on fast days. On the latter point, some ministers refused to baptize a child born on Sunday because of an old superstition that children were born on the same day of the week that they were

Earl E. Shelp (ed.), Sexuality and Medicine, Vol. II, 69–85.

conceived. One of the leading exponents of this view was the Reverend Mr. Loring of Sudbury, Massachusetts, who changed his view when his wife gave birth to twins on a Sunday ([7], pp. 9–14).

In theory Puritans strictly opposed sex outside of marriage, and they enacted harsh penalties against transgressors. Adultery was punishable by death, and fornication by whipping. Rape, however, was not considered as serious as it is today, and was placed in the same category as fornication. Despite the sternness of the laws, the Puritans recognized human weakness, and in any event they found the ordinances almost impossible to enforce. The problem was that two factors tended to increase the number of sexual offenses. First was the excessive number of single men without normal sex outlets, and second were the many servants who lived in the homes of their masters. The court records show case after case of men servants being found in the bedrooms of maid servants or daughters. Probably even more serious were the many serving maids who were made pregnant by male members of the family. A good many court cases involved maids bringing charges against their masters for sexual misconduct.

The sheer number of breaches in the sexual code in the colonial records explains in part the Puritan tolerance of them. Despite the many convictions for adultery, only three persons were actually executed; the majority of offenders were ordered to stand on the gallows and be branded. Yet even branding in most instances was purely symbolical. Fornication was considered a much lesser sin than adultery. Studies of church records reveal that a third or more of firstborn children were delivered within six or seven months, or less, of the marriage. In circumstances such as these, the parents were compelled to arise in church and publicly admit to committing fornication. This public confession was usually considered punishment enough, although from the number of young couples who were forced to admit their sinfulness one would scarcely judge it a deterrent. In explanation it should be noted that an engaged couple was considered married to all intents and purposes. Moreover, a young man frequently traveled miles to visit his bride-to-be, and the cold New England winters combined with the lack of heating encouraged the practice of bundling. This latter permitted the young couple to lie down together separated only by a narrow board or its equivalent [1].

The relaxed attitude towards young people and sex reflected in the 18th-century diary of the Reverend Ebenezer Parkman clearly belies the traditional picture of the Puritans. When he was married in 1724 he thanked God for "Restraining me from the Sin of Fornication and carrying me through so many Temptations as I have pass'd in ye Time of Courtship." Although he had resisted temptation himself, he was quite understanding of the failures of others. On his way to Hartford, Connecticut, in 1744 with a fellow minister, he recorded:

By Day we Wake wth ye Silent sight of a Young fellow in room getting up from his girl in ye t'other Bed in ye same Room with us – astonishing Boldnes and Impudence! Nor c[d] we let ye girl go off without a brief Lect.[e]; But we kept ye matter from ye Parents for ye time.

On another occasion he wrote: "On my way I went to Mr. Eleazer Bellows to admonish his Daughter Charity for her Repeated Fornication, w[ch] Duty I (as I was able) discharged and she thanked me for it" [34].

Colonial American society was largely rural and small town, and misconduct of any kind could scarcely be kept secret. As long as a couple married, the church and community evinced little concern, and even fornication, as long as it did not result in pregnancy, could be overlooked. Pregnancy was another matter, however, and here the community's main concern was in identifying the father and making sure he took responsibility for the child. To the Puritans sex was a necessity, and the only solution was marriage. Fortunately, the rural life and the availability of land made it possible for young couples to marry relatively early, thus relieving much of the sexual pressure.

The attitude of the Middle and Southern colonists was probably even more tolerant than that of the Puritans, but here again early marriages were common. In the South the existence of slavery, which gave white men access to slave women, gradually led to the custom of exalting white females and placing them on a pedestal. Elsewhere, as noted in New England, the presence of maid servants, many of them bonded, provided fair game for aggressive males. Occupying an inferior social status and completely dependent, they were subject to the whims of their masters.

Clearly, at all times and in all areas, there were many young men and women who adhered to the Christian injunction against sex before marriage, but the colonials took a casual attitude towards sex. One has only to read *The Secret Diary of William Byrd of Westover, 1717–1721* or a biography of Benjamin Franklin to appreciate that many in the upper classes had few sexual inhibitions. Byrd could casually pick up a prostitute in the streets of London, go home, say his prayers, and sleep with a clear conscience. His only worry seems to have been when he allowed his sexual activities to cause him to neglect his nightly prayers [38]. Even the virtuous Presbyterian minister, Philip Vickers Fithian, while describing Miss Betsy Lee in 1774 complained about her wearing the new fashionable stays which came "so high we can have scarce any view at all of the Ladies Snowy Bosoms" ([21], p. 172).

The Revolutionary and early national period witnessed the peak popularity of what are called the "Aristotle" books. These works described the sex organs in detail and expressed a very positive attitude towards sex. Sex was described as healthy for "It eases and lightens the body, clears the mind, comforts the head and senses, and expels melancholy." For men it was a necessity since "sometimes through the omission of this act dimness of sight doth ensue, and giddiness" ([32], pp. 27–28). In addition to the "Aristotle" books, which continued to appear until at least as late as 1831, many "facts-of-life" books and treatises dealing with venereal disease circulated widely.

In part because of a measure of openness towards sex and in part as a result of social and economic conditions, prostitution was a minor problem until the 19th century. The rural society, the shortage of women, the presence of slaves, indentured servants and, on the frontier, Indian women all tended to discourage prostitution. With an excess of marriageable males and without the economic pressure of an industrial system to push poor women into it, there was less occasion to enter prostitution. The latter, as with disease, is essentially a product of population density, and it emerged with the growth of urban areas. The New York newspapers in the 1750's mention arrests for prostitution, and there is evidence of it on a minor scale in other cities and towns, but it was not until well into the 19th century that prostitution became relatively common in America [16] [17].

The 19th century saw a radical change in America's attitude towards sex, and one of its more interesting aspects was the rapidity with which it occurred. Of the four works collectively referred to as the "Aristotle" books, the most popular one, *Aristotle's Masterpiece*, was published in at least 27 editions between 1766 and 1831. On this latter date, however, there appeared J. N. Bolles' *Solitary Vice Considered*, the first of a whole series of anti-masturbation tracts. In quick succession an American edition of Samuel A. A. Tissot's *Discourse on Onanism*, an early 18th-century work, was issued in 1832, a year later William A. Alcott included a chapter on sexual hygiene in his *Young Man's Guide*, and in

1834 Sylvester Graham published what was to become one of the most influential works of its day, *Lecture to Young Men on Chastity* ([32], pp. 24–28, 34). The contrast between the approach of the "Aristotle" books towards sex and the fear of sexual excess and masturbation expressed in these other works could scarcely be greater.

To complicate the picture further, in 1830 Robert Dale Owen's *Moral Physiology* was printed in the United States, the first American publication on birth control. Two years later the first American to write on the subject, a respected Massachusetts physician named Charles Knowlton, published his *Fruits of Philosophy, or the private companion of young married people*. Oddly enough, the birth control works of Owen and Knowlton went through many editions and were almost as widely read as those books expressing fear of sexuality ([32], pp. 24–28, 34; [3], pp. 220–21).

Thus between 1830 and 1834 there occurred the end of the "Aristotle" publications, the beginnings of birth control monographs, and the first of many writings expressing a fear of sex, which would eventually determine the public's attitude towards it. What may have proved decisive in shaping the course of American thinking was the intervention of the medical profession. As of the 1830's physicians could neither explain nor cure the great epidemic diseases and had virtually no understanding of the metabolic and degenerative disorders. Without an adequate basis to justify their position as a learned profession, they sought to emphasize their moral role in society. In consequence they happily seized on issues such as sexual morality and abortion, areas in which they claimed to have scientific knowledge, to bolster their standing in the community. By 1835 leading medical journals had picked up the themes of sexual excess and masturbation, and within a few more years the entire medical press was in full cry on the subject ([11], pp. 3–21).

Even granting that underlying factors had been in operation for sometime, it is rare that a major shift in public opinion can be pinpointed so closely. The rapidity of the change can be illustrated by noting that Knowlton was sent to jail in 1832 for publishing his views on birth control. The complete transformation in the attitude towards sex is further shown in the reaction by newspapers towards venereal diseases. Advertisements by physicians, or purported physicians, to cure venereal disease were common in newspaper columns, along with those promoting a host of proprietary medicines for "secret disorders." These advertisements, and those for abortionists, continued to fill advertising columns; but by the second half of the century as the subject of sex became taboo in polite circles, the public reaction towards it was expressed by an editorial in the Pittsburgh *Commercial* which announced that the paper would no longer print advertisements for cures of diseases such as syphilis, which were "brought on by vice . . . " [18]. As John Burnham wrote in 1972: "The most eloquent comment on the history of sex in the United States in the nineteenth century, however, is that the fullest and highest quality work in the field comes in accounts of attempts to suppress it!" [4].

Aside from the role of the medical profession, the best explanation for this sudden replacement of the acceptance of sex with the fear of it is to be found in the rapidly changing social order. Whereas at the beginning of the century America was essentially rural and self-sufficient, by the 1830's families were buying their food from bakers and food processors and their clothes and other goods from stores. Cities and towns were growing, and society was becoming increasingly interdependent. Sylvester Graham, one of the great reformers of his day, initially began his diet crusades by fighting for home-baked bread made with home-grown cereal. In the same sense that the natural food fad of today arose in reaction to the development of synthetics, chemical fertilizers, insecticides, and fast foods, so Graham and his contemporaries responded to a changing order. Graham started his career as a temperance worker, moved into the area of vegetarianism and diet, and, consistent with his physiological argument for a rigidly controlled diet, applied his thesis to sex. By the 1830's Graham had concluded, contrary to most medical thinking, which held that stimulus was essential to life, that stimulus was the cause of all physical disability. In particular, rich foods such as meat, pastries, condiments, coffee, tea, and wine overstimulated the body and led to such vices as masturbation and excessive venery, both of which quickly destroyed both the body and mind ([32], pp. 24–28, 34).

The advantage of attacking masturbation was that while the practice was almost universal, it had never been condoned in Christian writings; hence a measure of guilt was associated with it that was easily compounded by emphasizing its fatal physiological effect on the mind and body. Once Graham had classified masturbation as a physiological problem rather than a moral one, the medical profession quickly took over. What was probably the first medical article on the subject appeared in the *Boston Medical and Surgical Journal* in 1835 under the title, 'Remarks on Masturbation.' The author, who signed himself "W", explained that the habit was quite common among young people and was largely responsible for wasting their "vital energies." After listing such typical symptoms as feebleness and lack of vigor, he pointed out that a downcast countenance and shamefacedness revealed that the unfortunate victim "was conscious of his degraded condition. . . ." In the next issue of the *Journal* he described in detail how masturbation led to insanity. The victim of this practice, he asserted, "passes from one degree of imbecility to another, till all the powers of the system, mental, physical and moral, are blotted out forever" [36].

Not all physicians were willing to view sex askance, but the majority of correspondents and contributors to medical journals unhesitatingly accepted masturbation as a disease and recommended various forms of treatment, including castration and cauterization. In 1842 a physician made the startling discovery that "the habit of self-pollution" was widespread among females. Even more shocking, it was not restricted solely to the lower classes. As with many who wrote on the subject, he found some unusual symptoms associated with the disorder. By "diligent inquiry," he was able to cure three patients with chronic ophthalmia when he traced the cause of this symptom to their secret habit. Another physician successfully treated an 18-year-old Irish girl of hiccups when he earned she was a masturbator ([14], [24]). The prevailing medical opinion as to the danger from improper sexual practices was confirmed when the annual reports from insane asylums began to show the impact of masturbation on the mind. According to the Worcestor State Hospital *Report* in 1845, of the 2,300 cases of insanity admitted during the previous 12 years, some 145 resulted from masturbation. Three years later it was claimed that 32 per cent of those admitted to the Hospital were suffering the consequences of "self-pollution" ([23], p. 55).

By the mid-century those who argued that so-called unnatural sexual practices and even excessive sex within marriage were dangerous to health had carried the day. In 1849 the editor of the New York *Daily Tribune* lamented that the moral education of youth was being neglected "while the passions of a majority are prematurely developed and inflamed by stimulating food and drink, want of proper exercise, vacant hours, licentious books and corrupting associations" [15]. The pervasive influence of Sylvester Graham's dietary ideas was reflected in the practice of a Dr. John Walton who reported in the *American Journal of Medical Science* that he had successfully cured a case of nymphomania by placing his patient on a low protein diet [37].

By the mid-century, too, American male intellectuals were busily convincing themselves that women were inferior, delicate, sensitive creatures. This was a subject about which all professions could agree, but the medical profession, with its scientific knowledge of anatomy and physiology, was best positioned to supply the most convincing arguments. In 1847 Dr. Charles D. Meigs, a leading obstetrician and gynecologist, expressed what would remain the dominant opinion until well into the 20th century when he stated that a woman is "a moral, a sexual, a germiferous, gestative and parturient creature" [35]. This delicate, child-like being, subject largely to her emotions, under normal circumstances derived no satisfaction from sex other than that of pleasing her husband. In addition she was subject to a regular illness that periodically incapacitated her physically and mentally. Because of her highly sensitive nerves, her medical and emotional problems had been aggravated by the hurly-burly and turmoil of the advancing industrial society.

Intrigued by this major difference between the two sexes, the medical profession went to great lengths to explain the physiological and emotional makeup of women. Since the sexual drive was unnatural to females, any undue interest in sex – and masturbation in particular – was clearly pathological. Probably adding to this interest in females by male physicians were the sexual inhibitions and taboos of the late 19th century. Women were clothed from head to foot, and even obstetricians and gynecologists were not permitted to examine women visually. In 1850 a professor of midwifery at Buffalo Medical College permitted his students to witness a parturient patient. The result was a series of outraged editorials in newspapers and strong criticism even from some of the medical journals. The *New York Medical Gazette* that same year argued that virtually all procedures in obstetrics could be done "as well without the eye as with it." Skilled men could easily deal with even extraordinary obstetrical cases "by touch alone . . . " [13].

It was clear to the vast majority of Americans that women were unfitted for the professions, and this was especially true for medicine. Nineteenth-century medical journals are replete with editorials and letters from physicians either ridiculing the idea of women doctors or expressing shocked horror at the thought of women medical students seeing male bodies. Ironically, most women shared this view, even though the taboo against women revealing what were termed their 'sexual secrets' did irreparable harm to many of them. In a day when few doctors or midwives were competent and women endured repeated births and miscarriages, serious gynecological problems were common. Understandably, most women were reluctant to discuss their symptoms with a male physician and, for some, an examination was unthinkable. Jeannette Sumner, one of the pioneer physicians in Washington, D.C., reported in 1882 how she had saved the life of a woman with infected uterine tumors. The patient had been treated by male physicians for malaria because she had been too embarrassed to divulge her true symptoms [30].

An indirect benefit of the secrecy with respect to the female genitals was that it provided a major argument for permitting women to enter medicine. While most physicians agreed that no sensitive female could survive medical training without becoming coarsened, enough of them recognized the need for women physicians to treat certain female conditions that women were enabled to establish a foothold in the profession. This same secrecy aroused a measure of prurient interest and undoubtedly contributed to the veritable flood of articles on females in popular and professional journals in the post-Civil War years. Although it was clear that males were more prone to masturbation than females, most medical articles dealt with its practice by females. The psychological implications of this interest in an age of sexual repression are intriguing.

Two developments in the late 1860's intensified interest in female sexual proclivities. Dr. Isaac Baker Brown, an aggressive English surgeon, introduced the clitoridectomy, or excision of the clitoris, for the relief of epilepsy and nervous afflictions in females. A warm debate over the value of this procedure ensued in the English medical journals, with most authors questioning first whether masturbation was the underlying cause and second the usefulness of the operation [12]. The reaction in the United States was mixed, but the procedure gained some credence. The second development was the simultaneous discovery by physicians in France and England that women working in sewing machine factories were nervous, run-down, tired, and suffering from various other symptoms due "to immoral habits, . . . induced by the erethism which the movement of the legs evoked" [8]. Since the majority of these women were overworked, undernourished and poorly housed, it is not to be wondered that they exhibited signs of ill-health. And any physician who "discreetly" inquired about the sexual practices of a large number of individuals was bound to find ample evidences of what were considered "immoral habits"; thus 19th-century doctors had no trouble confirming their assumptions. The sewing machine cases led a good many physicians, whenever they were at a loss for a diagnosis, to make the same 'discreet' inquiries among their middle-class patients. It is not surprising, therefore, that medical journals were filled with reports of the dire results of irregular sexual practices nor that energetic surgeons, not content with excising the clitoris, began removing the ovaries and even the uterus in their determination to cure what they and the public called a "moral evil".

In the latter part of the century the interest in masturbation was obsessive. In the case of males it was assumed that the loss of sperm was destructive to the brain and nervous system, and this in turn created grave apprehensions about night emissions. For both boys and young men, elaborate devices were designed to prevent erections during the night or to restrain the hands. Among the more intriguing of these were the various toothed urethral rings. These were circular bands or rings with sharp teeth on the inside which were placed around the penis at night ([27], pp. 397–401). The phrase, "a rude awakening", is scarcely adequate to describe the effect of this contraption! While surgery was not uncommon on females, physicians were reluctant to castrate males. Nonetheless, occasionally they bit the bullet and resolutely did what was essential for their patients' well-being. Dr. J. M. Comens of New York, after trying all standard measures for curing masturbation in a young man, acceded to the patient's appeal for castration. He reported that since the operation the patient was "dull in intellect, sluggish in his movements," but he was a "strong healthy man!" [6].

By 1900 a few physicians were occasionally suggesting that masturbation was not too serious a problem, but by this time the doctrine was well established with the general public. With the advance of the 20th century the surgical procedures were gradually eliminated, but physicians continued to warn of mental breakdowns and grave physiological consequences of this practice. As late as 1922, Dr. Everett M. Ellison, a clinical professor at George Washington University Medical School, described a rather horrifying treatment given to a seven-year-old girl. The parents had discovered her masturbating at the age of three. Despite a variety of punishments and threats, which included beating her, placing her in a tub of water with her clothes on and threatening to drown her, the child continued. Realizing the need for drastic action, the physician had the child taken to the Emergency Hospital where she was shown scalpels and knives and warned she would have to undergo an extensive and mutilating operation which would leave her abnormal and deformed. When the child began "crying pitifully and begging for mercy," they agreed to postpone the operation. She had been watched closely since that time and was apparently cured. Reflecting a transitional viewpoint, Dr. Ellison asked at the end of his case report first if the child's masturbation would affect her mentally later in life, and second if the fright would eventually produce frigidity or mental abnormality [20].

While the Puritans did not condone the double standard of sexual morality, it was well embedded in Western society and applied to the rest of the American colonies. By the 19th century it was generally accepted everywhere. As justification for the double standard, it was argued that in a patriarchal society men could not commit infidelity since they introduced no new blood into the family. Moreover, a woman who engaged in sexual activities either before marriage or extramaritally would have a sense of guilt. As a result her health would suffer and her relations with her husband would be harmed.

One of the more intriguing questions of the 19th century was why women accepted their presumably passive role in sexual intercourse. The conspiracy of silence that had descended on all mention of sex or sex-related topics undoubtedly contributed to conditioning them to this role. Possibly more significant in determining their attitude towards sex was the suffering and death resulting from continued child-bearing. Prolapsed uteri, torn cervixes, and assorted fistulae were common occurrences and the intense pains of childbirth accompanied too often by hemorrhages, puerperal fever, and other complications understandably created a fear of sexual relations [35]. Women who had endured several pregnancies under the relatively primitive care of untrained midwives or poorly qualified physicians were not likely to stress the joy of sex to their daughters. Women's attitudes towards sex can clearly be seen in the Anti-Abortion Movement of the 19th century. The feminists, who might have been expected to support abortion and contraception, generally did not do so. A number of feminists blamed the rise of abortion entirely on the male sex. For example, in 1869 the *Woman's Advocate* of Dayton, Ohio, declared: 'Till men learn to check their sensualism, and leave their wives free to choose their periods of maternity, let us hear no more invectives against women for the destruction of prospective unwelcome children. . . ." As James C. Mohr and other writers have pointed out, most feminists saw the answer to unwanted pregnancies in abstinence ([29], pp. 111–113).

By more than a coincidence, most 19th-century articles dealing with sex appeared in medical journals. The topic was shunned in newspapers and magazines except for occasional grave warnings couched in euphemisms. Only physicians, who presumably considered the subject from a purely scientific standpoint, were permitted to write openly about it. Although discussions of the specifics of sex were restricted to medical journals, a wealth of material was published by ministers and popular lecturers warning young men and women of the grave danger arising from the sexual urge. This purity literature tended to emphasize the moral threat primarily, but the authors were not averse to embellishing medical arguments on the need for rigid control of sex.

As noted earlier, prostitution did not emerge as a serious social problem before the 1830's and 1840's. Newspaper and medical journals carried occasional complaints about it, but it was not until the second half of the 19th century when the medical profession gradually became aware of the serious nature of venereal diseases that the subject received much discussion. The editor of the *New Orleans Medical and Surgical Journal* spoke for most Americans in 1855 when, in the course of discussing the issue of controlling prostitutes, he pointed out that American women had a much higher morality than those of other countries and that most prostitutes in New Orleans were foreigners [19].

The Civil War undoubtedly contributed to an increase in prostitution and to spreading venereal diseases and by so doing helped create an awareness of the twin problem. In any event, in 1867 the Metropolitan Board of Health of New York City in its *Second Annual Report* estimated that the city had at least 20,000 cases of venereal disease, most of which it attributed to contact with prostitutes. Its Sanitary Committee on Prostitution recommended a strong program of regulation, which included registering all brothels and their occupants, providing medical inspection, and establishing a hospital to treat infected prostitutes. Realizing that in mentioning the subject they were treading on dangerous ground, the Committee members wrote that those who judge harshly should "reflect that if the Creator has affixed penalties to vices, He has, in His wisdom and goodness, furnished us with remedies to treat the diseases which follow them. . . ." A major recommendation of the Committee was that all hospitals and dispensaries receiving state aid must be required to accept venereal disease patients [26]. Many hospitals, even into the 20th century, refused to admit those whose complaints were the wages of sin, and many physicians would not treat them.

To regulate vice, however, is to condone it, and New Yorkers were not willing to acknowledge the presence of prostitution or venereal disease. Three years later St. Louis became the first American city to set up a regular system of medical inspection for prostitutes. The City Council, on the recommendation of the City Health Officer, enacted on July 5, 1870, what was termed the "Social Evil Ordinance." This law divided the city into six districts with a physician assigned to each one. These medical inspectors were to examine all prostitutes in their areas and to send those with infections to a special Social Evil Hospital. Although individual physicians supplied leadership in the movement to regulate prostitutes, the medical society in St. Louis was divided on the question, with at least a third or more of the members objecting to legalizing sin [5].

The Social Evil Ordinance lasted only four years, but the St. Louis experiment touched off a national debate on the subject. The American medical profession was divided into those who favored inspection, those who doubted its effectiveness, and those who opposed inspection on moral grounds [5]. Laymen were generally opposed, and, in part as a response to the agitation for medical inspection, a purity movement developed with the aim of abolishing prostitution. This movement gained strength in the early 20th century when a group of concerned physicians, apprehensive over the spread of venereal disease, organized the social hygiene movement. Despite their joint efforts, it was not until basic changes occurred in economic and social conditions combined with a new moral standard that the traditional red light districts were eliminated and the incidence of prostitution reduced.

The struggle by physicians and public health leaders to make the public aware of the seriousness of venereal disorders is itself a commentary on the prevailing attitude towards sex. When the American Public Health Association considered taking a public stand against venereal diseases in 1882, they were dissuaded on the grounds that the membership included laymen who should not even hear discussions on a topic that was "purely a medical one. . . ." Howard Kelly, one of the leading medical professors at Johns Hopkins, in 1899 opposed a discussion in the AMA on "the hygiene of the sexual act" because it would be "attended with filth and we [should] besmirch ourselves by discussing it in public" [9]. The New York Academy of Medicine ran into similar problems when in 1892 papers were presented on how to prevent the spread of syphilis. A suggestion to educate the public through newspapers brought outraged protests from two of the members ([28], pp. 31–41).

By 1900 the climate of opinion was changing, and medical journals began carrying more articles and editorials on venereal diseases. The man who deserved much of the credit for removing the veil of silence was Dr. Prince A. Morrow of New York. He took the lead in bringing the subject before medical groups and later moved into the public arena. Despite his efforts, the New York City Health Department in 1912 ran into a storm of protest when it required a limited reporting of venereal disease cases [2]. Apprehension over the spread of these infections during World War I led the United States Public Health Service to establish a Division of Venereal Disease. Sensitive to public opinion, no mention was made of venereal disease in print, and the law creating the Division was officially entitled "the Army appropriation bill" ([22], p. 320).

The 1920's brought a revolution in morals and culture, the precise reasons for which are not clear. One major factor was World War I, since it created a generation gap and at the same time turned the United States into a major industrial nation. In any event, the role of women changed drastically, and the image of the sexless frail female rapidly faded. The delicate young woman who could not appear in public without a chaperon or protector was replaced by the working girl, the flapper, and the business woman. Among the freedoms they demanded were those of a sexual nature. Sigmund Freud, whose works had elicited little public notice prior to the war, now became a national rage. Misreading his works, popular writers glibly used his terminology to justify, at least by implication, sexual freedom. Newspaper tabloids, motion pictures, and advertising agents quickly discovered that sex would sell, and the exploitation of sex became a major theme in popular culture.

The graphic nature of sexual exploitation by the motion picture industry led most of the states by the 1920's to enact some form of censorship. In reaction, the industry decided to police itself and in 1922 appointed Will Hays, Harding's Postmaster-General, to do the job. As

William Leuchtenburg has observed, the result was to add hypocrisy to sex by insisting on false moralizations ([25], pp. 209–220). To what degree sexual promiscuity increased in the 1920's is difficult to determine, but there was much greater sexual freedom among the middle and upper classes. Undoubtedly wider knowledge and improved techniques of birth control contributed to this change.

The Depression years of the 1930's brought a reaction to the excesses of the previous decade, but fundamental changes had been made. Nonetheless, whatever Americans may have done about sex, it was still dealt with gingerly in the media. No matter how sex was exploited, virtue – equated with chastity – had to be rewarded and immorality punished. When Thomas Parran, Roosevelt's Surgeon General, began the first major campaign against venereal disease in the mid-1930's, he encountered considerable opposition. For example, before giving his first radio talk on the subject, he was told by the station manager that if one call was received objecting to the subject, he would be cut off the air [10].

As might be expected, World War II led to some relaxation of sexual taboos, but the wide gap between what Americans professed about sex and what they did narrowed only slightly. The second sexual revolution began in the 1960's. By this time antibiotics had seemingly paved the way to eliminate venereal diseases, the pill and other methods of birth control had largely removed the fear of pregnancy, and a young generation emerged which had been reared in affluence. Largely free from economic worries, they were first politicalized by country singers such as Joan Baez and Pete Seeger, introduced to drugs by Timothy Leary and his disciples, and then turned on to sex by rock groups. In these years the cult of the natural appeared, one which mixed hedonism with opposition to conventional society. This cult of naturalism was quickly applied to sex, and sexual taboos began to disappear. Women's clothing became scantier, topless and bottomless became common ploys to attract bar customers, and sexual minority groups began coming out of closets. Where formerly homosexuals and lesbians had been objects of legal harassment and public scorn, they now began to demand their rights as citizens and respect as human beings. Within a few years it became acceptable for unmarried couples to live together, for single women to have children, and for sexual minorities to participate in state and church institutions ([33], pp. 233–271).

None of this was accomplished without a considerable backlash.

Ironically, the middle class, which had sought to impose its impossible standards on America in the 19th century, led the sexual revolution in the 20th, and the working class, which had paid only limited attention to 19th-century sexual standards, resisted the new openness. The public's reluctance to permit the teaching of sexual hygiene in schools indicates that the ideas of the intellectuals have still not permeated very deeply into society.

University of Maryland,
College Park, Maryland, U.S.A.

BIBLIOGRAPHY

1. Adams, C. F.: 1891, 'Some Phases of Sexual Morality and Church Discipline in Colonial New England', Massachusetts Historical Society *Proceedings*, second series, VI, 503–509.
2. Biggs, H. M.: 1913, 'Venereal Disease', *New York Medical Journal* **97**, 1009–1012.
3. Bullough, V. L. and B. L.: 1964, *The History of Prostitution*, University Books, New York.
4. Burnham, J. C.: 1972, 'American Historians and the Subject of Sex', *Societas* 2, 309.
5. Burnham, J. C.: 1971, 'Medical Inspection of Prostitutes in America in the 19th Century: The St. Louis Experiment and Its Sequel', *Bulletin of the History of Medicine* **45**, 203–218.
6. Comens, J. M.: 1869–1870, 'Onanism Among Children', *Transactions of the Eclectic Medical Society of New York* **4**, 755–670.
7. Degler, C. N.: 1970, *Out of Our Past, The Forces That Shaped Modern America*, Harper & Row, Publishers, New York.
8. Down, J. L. H.: 1867, 'Influence of the Sewing Machine on Female Health', *New Orleans Medical and Surgical Journal* **20**, 359–360.
9. Duffy, J.: 1979, 'The American Medical Profession and Public Health: From Support to Ambivalence', *Bulletin of the History of Medicine* **53**, 1–22.
10. Duffy, J.: 1961, Personal conversation with Dr. Thomas Parran.
11. Duffy, J.: 1982, 'The Physician as a Moral Force in American History', in W. B. Bondeson, *et al.* (eds.), *New Knowledge in the Biomedical Sciences*, D. Reidel Publishing Co., Dordrecht, Boston, and London, pp. 3–21.
12. Editorial: 1867, 'Clitoridectomy', *Southern Journal of Medical Science* **1**, 794.
13. Editorial: 1850, 'Demonstrative Midwifery', *New York Medical Gazette* **1**, No. 11, 166–168, No. 13, 106–107.
14. Editorial: 1842, 'Masturbation', *Boston Medical and Surgical Journal* **XXVII**, 102–108.
15. Editorial: 1849, *New York Daily Tribune*, January 20.
16. Editorial: 1755, *New-York Gazette or the Weekly Post-Boy*, April 11.
17. Editorial: 1756, *New-York Gazette or the Weekly Post-Boy*, May 31.
18. Editorial: 1867, Pittsburg *Commercial*, June 12.

19. Editorial: 1855, 'Review of European Legislation for the Control of Prostitution', *New Orleans Medical and Surgical Journal* **11**, 700–702.
20. Ellison, E. M.: 1922, 'A Case of Long-Continued Masturbation in a Girl, Cured by Fright', *New Orleans Medical and Surgical Journal* **74**, 160–165.
21. Farish, H. D. (ed.): 1943, *Journal & Letters of Peter Fithian, A Planter Tutor of the Old Dominion*, Colonial Williamsburg, Inc., Williamsburg.
22. Furman, B.: [n.d.] *A Profile of the United States Public Health Service, 1798–1848*, United States Government Printing Office.
23. Grob, G. N.: 1966, *The State and the Mentally Ill*, University of North Carolina Press, Chapel Hill.
24. Letter to the Editor: 1845, *Boston Medical and Surgical Journal* **XXXII**, 195–197.
25. Leuchtenburg, W. E.: 1978, 'The Revolution in Morals', in J. H. Cary and J. Weinberg (eds.), *The Social Fabric*, Little, Brown and Company, Boston and Toronto, pp. 209–220.
26. Metropolitan Board of Health of the State of New York: 1867, *Second Annual Report*, 31.
27. Milton, J. L.: 1887, *On the Pathology and Treatment and Gonorrhoea and Spermatorrhoea*, New York, pp. 397–401.
28. Minutes of the Section on Public Health; May 17, 1892, New York Academy of Medicine.
29. Mohr, J. C.: 1978, *Abortion in America: The Origins and Evolution of National Policy*, Oxford University Press, New York.
30. Moldow, Gloria M.: 1980, *The Gilded Age, Promise and Disillusionment: Women Doctors and the Emergence of the Professional Middle Class in Washington, D.C.*, Ph.D. Dissertation, University of Maryland, 139–140.
31. Morgan, E. S.: 1942, 'The Puritan and Sex', *New England Quarterly* **15**, 591–607.
32. Nissenbaum, S.: 1980, *Sex, Diet, and Debility in Jacksonian America*, Greenwood Press, Westport.
33. O'Neill, W. L.: 1971, *Coming Apart, An Informal History of America in the 1960's*, Quadrangle Books, Chicago.
34. Parkman, E.: July 7, 1724, February 15, 1744, Diary, American Antiquarian Society ms.
35. Smith-Rosenburg, C. and Rosenburg, C.: 1973, 'The Female Animal: Medical and Biological Views of Woman and Her Role in Nineteenth-Century America', *Journal of American History* **60**,332–356.
36. 'W': 1835, 'Remarks on Masturbation', *Boston Medical and Surgical Journal* **XII**, 94–97; *ibid.*, 'Insanity Produced by Masturbation', 109–111.
37. Walton, J. T.: 1857, 'Case of Nymphomania Successfully Treated', *American Journal of Medical Science*, n.s., **33**, 47–50.
38. Wright, L. B. and Tinling, M. (eds.): 1958, *William Byrd of Virginia, The London Diary (1717–1721) and Other Writings*, Oxford University Press, New York.

ROBERT BAKER

THE CLINICIAN AS SEXUAL PHILOSOPHER

> A great sexual sermon – which has its subtle theologians and its popular voices – has swept through our societies in the last decades; it has chastised the old order, denounced hypocrisy, and praised the rights of the immediate and the real; it has made people dream of a New City. . . . And we might wonder how it is possible that the lyricism and religiosity that have long accompanied the revolutionary project have, in Western industrial societies, been largely carried over to sex.
>
> Michel Foucault ([8], pp. 7–8)

VALUES AND THE CLINICAL CONTEXT

The clinic is a world of sex and gender, of males and females, of masculinity and femininity, of typically male/masculine doctors and typically female/feminine nurses. These divisions have reverberations that are keenly felt throughout medical practice. In ancient China, for example, respectable females used special feminine dolls to explain their symptoms to their invariably male doctors; in contemporary America, typically masculine physicians take great care to drape their invariably female patients for gynecological examinations. The ancient doll and the contemporary draping both reflect adaptations evolved by the clinical world to cope with the sexual philosophy of the wider society. Since these adaptations are extremely revealing, some might see in them a type of unconscious philosophical analysis – yet when I characterize clinicians as sexual philosophers in the title of this essay, I intend nothing as oblique or esoteric as this.

Earl E. Shelp (ed.), Sexuality and Medicine, Vol. II, 87–109.

Nor does this essay directly address the tendency to conceptualize sexual matters in medical terms. To be sure, conceptions of sexual practice are malleable. One and the same sexual practice can be conceptualized in different ways. Thus in classical Greece, pederasty and homosexuality were often regarded as elements of a sophisticated upper-class life-style. In the Middle Ages, however, the very same activities were thought of as sinful perversions ([2], [7]). The sinfulness of these forms of sex was "medicalized" in nineteenth and twentieth centuries, when clinicians taught the world to think of these "perversions" as "medical disorders" ([1], pp. 370–420; [4], pp. 333–352, 693–724). The conceptual odyssey of these practices – from upper-class life-style, to sinful perversions, to psychological disorders – is inherently philosophical and one could, if one wished, suggest that those clinicians responsible for these conceptual innovations are engaging in a form of sexual philosophy. Yet this is not what is meant when clinicians are referred to as "sexual philosophers" in the title of this essay.

What, then, is this essay about? It is about the values and theories that clinicians draw on when they advise their patients about sexual matters. For – in the clinical practice of medicine – the voices of contemporary clinicians preach what French philosopher-historian Michel Foucault refers to as the "great sexual sermon" – chastising the old order, denouncing hypocrisy, and praising the rights of the immediate and the real. Some might protest that when competent clinicians discuss sexual matters with their patients, they do so from a medical, therapeutic, or scientific perspective – they do not discuss these issues philosophically. Such a response reflects a common misperception of the nature of philosophical theory. No one wishes to challenge the proposition that, for the most part, when clinicians discuss sexual matters with their patients, they attempt to do so in a manner that draws appropriately on their scientific knowledge and therapeutic training. The point to be taken is that when sexual events conspire to force patients to engage their clinicians in discussions of sexual matters, the discussion necessarily evokes philosophies of sex *not* because clinicians act in an unscientific or non-therapeutic manner, but rather because both science and therapy presuppose normative and philosophical commitments. Admittedly, these discussions may appear to be philosophically neutral to the participants, but that, as I shall show, is precisely the reason that these discussions become morally problematic.

To prove this point, I ask the reader to reflect on an apparently simple case. The case is drawn from a series assembled by Dr. Richard Cross as teaching material for the clinicians and pre-clinical students at the Rutgers Medical College ([1], p. 460).

MATERNAL WORRIES ABOUT MASTURBATION

A family practitioner reported that a thirty-eight-year-old, college-educated woman, the wife of a prominent businessman, was concerned about her fourteen-year-old son, the eldest of her children, Fred.

"I probably shouldn't be bothering you, Doctor, but I am concerned about Fred, and I don't know where to go for advice. I know masturbation is perfectly normal, but really, he is overdoing it. Of course, he is no longer a baby, but he is only just fourteen, and almost every night there are fresh stains on his sheets, and his pajamas are a mess. I'm ashamed to send his things to the laundry, and I can't get all the stains out.

"Alex (that's my husband) tried to convince me that I was accusing him unjustly and that he was just having wet dreams, but I know he's guilty because last week I caught him in the act. I was sewing in my bedroom in the late afternoon when I heard noises coming from his room. He had a book in front of him and the next day while he was in school I found this [book, *Fanny Hill*] hidden in his bottom drawer. It is disgusting!

"He's a fine boy, Doctor, and I just want to be sure he's not being led astray. He's on the football team in Junior High, and he goes with a pretty tough bunch of kids, and I think they must be responsible. Alex won't do anything; he just says it's normal and not to worry. But I can't help worrying, Doctor."

On Seeking Advice

What kind of help is a clinician inclined to offer in this sort of case? Not surprisingly, the clinicians and students with whom I have discussed this case almost invariably echo the information and opinions given in the standard textbooks on human sexuality. Thus they stress the following two points:

(1) *Normality* – masturbation is "normal" in the sense that (to quote the oft cited Kinsey statistics) by the age of 45, 58% of all females and 92% of all males have masturbated to the point of orgasm ([12], p. 284, [9], p. 122);
(2) *Harmlessness* – contrary to previously held beliefs, "the physiological harmlessness of masturbation has been established" ([12], p. 267); "Masturbation is convenient, free, safe, and devoid of social or interpersonal difficulties" ([9], p. 116).

From these two facts they, like the human sexuality textbooks they parrot, draw two conclusions.

Conclusion One – negative attitudes towards masturbation "reflect the discredited but not discarded notions of former times" ([12], p. 284).
Conclusion Two – that for many people (particularly adolescents and nonorgasmic women) masturbation is extremely beneficial ([9], p. 117; [12], p. 284; [14], p. 107).

Finally, most discussants suggest that the appropriate response of the clinician in a case such as this is to begin to treat the mother – i.e., they would not only reassure her about the "normality and naturalness" of adolescent male masturbation, they would take this opportunity to explore her feelings, not only about Fred's masturbation, but about her own sexuality. They would attempt to defuse her feelings of anxiety about Fred and, if possible, to infuse her with a more enlightened, "healthier" attitude towards masturbation. The one point that medical discussants never seriously consider is that Fred's masturbation might actually be morally problematic. In hours of discussing this case, I have never heard a clinician or pre-clinical student seriously consider the possibility that this woman's reaction to her son's behavior might be legitimate.

The Valuational Presuppositions of the Clinical Context

One of the characteristics of a naively held philosophical theory is that it preempts the possibility of alternatives. The world is as it is conceived to be and, to the mind of the naive conceiver, cannot be imagined to be otherwise. One primary function of the professional philosopher is to create conceptual space. Consider again the two facts urged in favor of masturbation – that it is normal and harmless. Suppose that tomorrow 58% of all females and 92% of all males reaching their 10th birthday were incapable of growing over four feet three inches in height. Thus, tomorrow, four foot three inches becomes the norm for all humans who had been born ten years ago or earlier. Suppose further that this radical dwarfing of the human race was otherwise harmless. Would we not still regard this as a plague? Would we not seek a cure? And in seeking a cure would we not be admitting rather forthrightly that our aesthetic values infiltrate the concept of medicine? For surely, if dwarfing affected neither longevity nor other physiological functions, if the new generation of dwarfs were as functional as contemporary pygmies (or, to reverse the example, as Kareem Abdul-Jabbar), then, since these dwarfs were normal for their generation, there should be no grounds for therapeutic intervention.

But, of course, there would be innumerable attempts at intervention – not because the dwarfism would threaten lives or functionality, but rather because the physiology of these children would not conform to the aesthetic values of their parents.

Aesthetic, social, and moral values structure the clinical world, not merely in matters of height and sexual behavior, but universally. It could not be otherwise; for the essential difference between clinical medicine and such studies as zoology and pathology is that the clinical perspective is informed by a positive valuation of human comfort and survival, while the pure sciences are not. As one analyst, Peter Sedgwick, puts this point

> Human beings, like all other naturally occurring structures, are characterized by a variety of inbuilt limitations or liabilities, any of which may (given the presence of further stressful circumstances) lead to the weakening or collapse of the organism. . . . Out of his anthropocentric self-interest, man has chosen to consider as "illnesses" or "diseases" those natural circumstances which precipitate the death (or the failure to function according to certain values) of a limited number of biological species; man himself, his pets, and other cherished livestock, and the plant varieties he cultivates for gain or pleasure. Around these select areas of structural failure man creates . . . specialized combat-institutions for the control and cure of "disease": the different branches of the medical and nursing professions, veterinary doctors, and the botanical specialists in plant disease. . . . Plant diseases may strike at tulips, turnips, or such prized features of the natural landscape as elm trees, but if some plant species in which man had no interest (a desert grass, let us say) were to be attacked by a fungus or parasite, we should speak not of a disease but of a competition between two species. The medical enterprise is from its inception value loaded . . . ([4], pp. 121–122).

Usually the values loaded into medicine are unremarkable to those engaged in the enterprise, because they are shared by everyone – including patients. Thus both clinicians and patients value life over death, and prefer pleasure to pain and discomfort, etc. So, although the medical enterprise is shot through with values, they have become "transparent" through universality – like a pane of glass, one sees through them without noticing them. Values can become opaque and noticeable to the participants – but only in those instances in which the valuational consensus breaks down. Sometimes, however, there is a breakdown of consensus without a concomitant opacity of values.

Consider again the case of Fred's mother. Her values are those of a prior age. Thus, had she related her problem to a nineteenth century clinician he would, no doubt, have *agreed* that Fred's masturbatory behavior was filthy and unhealthy (see Benfield on the spermatic economy [2]) and proceeded to recommend one of the therapies in the medical armamentarium of the period – possibly including surgery, acid burns, and thermoelectrocautery ([4], pp. 267–280), or clitoridectomies, circumcision, and "female castration," had this been a case of female masturbation ([2], pp. 120–132). We think of thermoelectrocautery of the penis and wonder at the values that demand this treatment. But we see the treatment from the perspective of a valuational revolution that reversed those of the nineteenth century, a revolution that transformed masturbation from a pathology into a sign of normal, healthy development. Our values clash with theirs and so the issue of values becomes visible to all. In the case of Fred's mother, the same clash of values occurs. She subscribes to prerevolutionary values. Unfortunately, clinicians fail to recognize the role that value differences play in this dissent and respond to her inappropriately by citing statistics – as if these facts could somehow resolve a difference in values.

Limitations of the Clinician's Role as Valuational Vicar

Once the discrepancy between the values of Fred's mother and her clinician becomes apparent, it becomes clear that the more fundamental question posed by the case concerns the treatment of dissenters: how should clinicians deal with patients who do not subscribe to the prevailing valuational consensus in sexual matters? Once more, it is helpful to reflect on clinical encounters from previous eras. Consider the following scene from A. J. Cronin's autobiographical novel, *The Citadel*:

> In the spring of 1926 the good Edwal, newly married, had sidled, late, into Manson's surgery with the air thoroughly Christian, yet ingratiatingly man of the world.
>
> "How are you, Doctor Manson! I just happen to be passing. As a rule I attend with Doctor Oxborrow, he's one of my flock you know. . . . But you're a very up-to-date doctor by all accounts and purposes. You're in the way of knowing everything that is new. And I'd be glad – mind you I'll pay you a nice little fee too – if you could advise me." Edwal masked a faint priestly blush by show of worldly candor. "You see the wife and I don't want any children for a while yet anyhow, my stipend being what it is, like . . ."
>
> Manson considered the minister of Sinai in cold distaste. He said carefully:

"Don't you realize there are people with a quarter of your stipend who would give their right hand to have children? What did you get married for?" His anger rose to a sudden white heat. "Get out – quick – you dirty little man of God!"

With a queer twist to his face, Parry had slunk out ([6], p. 176).

What legitimated Dr. Manson's outburst? Dr. Cronin, or his surrogate, Dr. Manson, would, no doubt, have appealed to the societal values of the day. Prior to the Second World War, society accepted the normalcy of procreation and abhorred contraception (and yet, quite inconsistently, it honored abstinence – even though sexual abstinence is as inimical to procreation as contraception). Thus, viewed from the perspective of the value system of the period, Doctor Manson represented his society's positive valuation of procreation and life. Edwal Parry, the minister, was anti-life. His request contravened the values of society, medicine, and church, and was, therefore, an unnatural request for assistance in an unnatural act, which, as vicar for society, Dr. Manson properly chastised.

The contrast between the clinician, the true vicar of the values of twentieth-century society, and the Reverend, a false defender of these values, is a brilliant literary conceit, but does the scene play as a moral drama? Should clinicians use their clinics to actively proselytize for the prevailing societal values? Does Dr. Manson's medical license give him license to disparage any unpopular values he encounters in his practice? Does it permit him to dismiss a request for birth control information with the angry remark that the requesting patient is "dirty"?

The answer to all of these questions is negative. From Hippocratic times to the present, the ethic underlying all Western medical practice is the ethic of helping. The two most famous *dicta* associated with the Hippocratic corpus are injunctions to help the sick or, failing that, to do no harm ([5], pp. 14, 15). Did Dr. Manson help Reverend Parry when he refused him contraceptive information? Dr. Manson was clearly doing what the society of the time believed was right and proper, he was acting as he himself thought was appropriate, but I do not believe that anyone would argue that he helped the Reverend. Dr. Manson may have believed the Reverend's values misguided – witness the statement, "Don't you realize that there are people with a quarter of your stipend who would give their right hand to have children?" – but in attempting to convert Parry to the prevailing societal consensus, Manson exceeded his warrant as a physician.

The parallel between the Parry case and that of Fred's mother should be obvious. Neither accepts the prevailing societal consensus about sex. Both turn to their clinicians for advice. But, instead of finding an atmosphere open to the exploration of alternative values, they encounter clinicians who, by argument, by authority, by suggestion, and by innuendo, preach a valuational sermon – sometimes in the guise of a scientific discourse. The purpose of the sermon, in both cases, is to convert a nonconformist to the prevailing view of sex.

ALTERNATIVE VALUES

Alternative Conceptions of Masturbation

Clinicians are inclined to deny the parallel between their response to the case of Fred's mother and the Manson-Parry case. They fail to see the similarity between the cases because they neither accord Fred's mother's response philosophical legitimacy nor recognize the philosophical presuppositions underlying their own proclamations of normality. Part Two of this essay is an attempt to correct both of these misconceptions. It is dedicated to an exploration of the philosophical presuppositions, the value structures, of both clinician and patient.

Fred's mother may have been repelled by her son's masturbatory acts for many reasons. Perhaps she was a puritan. Perhaps she had spent an eternity of Sundays listening to St. Augustine's thoughts echoing in the sermons of Protestant ministers. Then again, she might have been ashamed of Fred's masturbation because she had learned the sexual rationalism of St. Thomas Aquinas, perhaps while sitting at the feet of Catholic nuns. Again, she had been infused with neo-Kantian sexual romanticism via *Love Story* or any number of popular films or novels. Not knowing any more about the woman or the case than the reader, I cannot tell which of these routes led her to consult her doctor about Fred. Not that the woman is likely to have read any of these philosophers, or to know their names. (Philosophy, although the most influential of academic disciplines, is the most arcane and the least well-known.) Thus Fred's mother was probably unaware of the role that St. Augustine, or St. Thomas Aquinas, or Immanuel Kant had played in structuring her world. But in all likelihood one of these philosophers is responsible for leading her to hold values that differ from the prevailing societal consensus. In the next three sections I shall explore these views so that they can then be compared with the more popular positions espoused by clinicians.

The Puritanism of St. Augustine

As we remarked above, Fred's mother may well have been a puritan. This description is not intended as a pejorative epithet. "Philosophical puritanism" is not a snide description ascribed to those whose views on sex are less "progressive" than our own. It is a view of sexuality that, although closely identified with a number of Christian sects – including the Puritans – actually predates Christianity. The puritanical elements in Christian thought, although firmly grounded in St. Paul's Epistles, might never have become so prominent had it not been for the work of the Bishop of Hippo, St. Augustine (A.D. 354–430). For Augustine and, indeed, for all philosophical puritans, puritanism has little to do with sexuality per se; it really concerns the proper relationship between mind/soul and body. At the heart of the theory lies the perception that one's moral stature and value as a human being is a function of the mind's ability to control the body.

To understand the puritanical vision, it is helpful to meditate upon some paradigmatically bestial creature – for example, a crab. Crabs lie at the opposite end of the evolutionary ladder from humans – even vegetarians are ambivalent about the moral standing of crabs. It might, perhaps, be morally problematic to eat a cow, a lamb, or a pig, but it is difficult to wax eloquent on behalf of the rights of crabs. Yet crabs are something more than vegetables. They move, they eat. They are clearly animals, pure beasts, as it were, without a hint of (morally redeeming) humanity in their nature. They do, however, exhibit rationally purposive behavior; for example, they catch and eat prey. And when a fiddler crab eats, whatever rational capacity the creature has would appear to be at the service of its bodily hunger. Bodily imperatives control the will of the creature, its psychological being – if, indeed, it has such a being – would seem to be little more than a mechanism for the satisfaction of its bodily appetites.

The puritanical world view is built on the perception that for a human being – for example, a human fiddler – the relationship between body and mind is the inverse of that illustrated by the case of the fiddler crab. For the human fiddler, the will is not an instrument of the fiddler's body; rather, the body is an instrument of the fiddler's will – in the case in point, the body becomes an instrument of the human will to create music. (For humans, when they are acting most human, the body is an instrument of the will; for animals, when they are most animal-like, the rational will is an instrument of bodily appetites.) For puritans, we exist as humans only when our rational and spiritual selves command our bodies; when the demands of the body command the will, we relinquish our humanity and become mere animate creatures. Thus, to quote Augustine, for humans the body "is subject to the mind because of its lower nature" ([15], p. 110).

The fundamental principle of puritanism is that humans are uniquely "unnatural" animals; only humans have the capacity, through their will, to transcend the dictates of their bodies. It is not this fundamental principle, however, that gives puritanism its "puritanical" reputation. The anti-libidinous aspects of the theory derive from the following line of reasoning: since humans are human essentially because the rational mind controls the body, insofar as humans permit their bodies to dominate their wills (e.g., insofar as they succumb to "lust"), they debase (pervert, invert) their humanity. Whenever appetite dominates humans, they debase themselves to the level of crabs – and like crabs they lose their moral stature. The vices of gluttony and vanity, like sexuality, all reflect an inability of the will to control the appetites of the body. They are vices precisely because they invert, and thereby degrade, our human nature.

Masturbation is a paradigm of the perverse domination of body over will. By masturbating we submerge our selves in our bodies to serve their ends, so that they dominate us. We do not control our penises, our vaginas; they control us; in orgasm we lose self control. As Augustine puts the point, had "human beings . . . remained innocent . . . they would have used their genital organs for the procreation of offspring in the same way that they use the rest, that is, *at the discretion of the will*" ([15], p. 112, italics added). To repeat, Augustine's point, indeed, the point for all puritans, is that in "lust" our sexual organs are unresponsive to our will, but our will responds to our organs.

If all of this rather high-sounding rhetoric seems incongruous in the context of an action as mundane and commonplace as masturbation, recall that our purpose in considering Augustine was to analyze the possible philosophical presuppositions that might have influenced Fred's mother's reaction to the sight of his masturbatory activities. As she reported the incident, she recalled hearing "rhythmic noises coming from Freddy's room," she quietly opened the door, and "there he was, so engrossed he never saw [her]." In other, more Augustinian, words, her son's body with its noises so dominated his will that he was oblivious to anything but his body. When she watched him she felt that she lost her son, the person she had raised, to a body. She was ashamed of, and disgusted by, this submission to his body. It is not merely that Fred had lowered himself to the level of a creature; it is rather that, unlike a crab, he had a choice of being human, and, momentarily at least, he chose to lower himself to the level of a mere creature.

It is altogether too easy to mock reactions like those of Fred's mother. Yet, as Augustine pointed out to his contemporaries, there is a measure of hypocrisy in the scorn with which sophisticates regard puritans. The sophisticates of Augustine's day were philosophical naturalists known as Cynics (apparently because they believed that humans should act in accordance with their animal nature – as canines do; "cynos" is Greek for "dog"). The Cynics are just as skeptical (cynical, if you will) about the puritanical analysis of sex as are contemporary "progressives." Augustine pointed out, however, that although the Cynics talked about enlightened attitudes towards sexual acts, although they claimed that "no one should be ashamed to do it openly and engage in marital intercourse in any street or square" ([15], p. 105), after a single public performance by Diogenes, the founder of the movement, "the later Cynics . . . abandoned the practice, and modesty prevailed over error" ([15], p. 105).

Contemporary "progressive" thinkers, like the ancient Cynics, tend to behave in a manner more modest than their rhetoric would suggest. Unlike cats, dogs, and other domesticated animals, humans tend to feel ashamed of public acts of copulation or masturbation. We are, in fact, singularly reticent about our masturbatory habits – no one publishes articles about the recent innovations in masturbatory technique in the popular press, people do not go about extolling the virtues of their new dildo or vibrator. On the contrary, we tend to treat masturbation as a shameful form of incontinence, rather like bed-wetting. We act, in fact, just as puritans would expect that we would. We admire athletes, ballerinas, and other superlative exemplars of the human control of body by will, and we admire least (and/or pity most) those whose bodies control their will – the physically sick, the weak-willed (e.g., the obese, the "out-of-shape"), and the incontinent: the bed-wetters and the masturbators. Thus, in fact, our attitude towards masturbation is rather more puritanical than we like to admit. We tend to be closet puritans, our public disparagement of the position merely disguises our inner inclinations to accept it.

The Sexual Relationalism of St. Thomas Aquinas

Fred's mother might not be a puritan, closet or otherwise; she might, instead, be a Thomistic rationalist. St. Thomas Aquinas (1225–1274) was the great reconciler. In his philosophical and theological writings he reconciled mind and body, reason and revelation, and the blood of Christ with a cup of wine. Thomas found it inconceivable that a beneficent Creator could have fashioned a world in which mind and body were eternal adversaries. He rejected the puritanical vision of human nature as a battleground between will and body. He saw the person as a whole consisting of parts, including the person's mind and body. The person, and each of the person's parts, has ends (or goals). These ends determine what is good both for the person and for the parts of the person.

> . . . God exercises care over every person on the basis of what is good for him. Now it is good for each person to attain his end, whereas it is bad for him to serve away from his proper end. Now this should be considered applicable to the parts, just as to the whole being; for instance, each and every part of a man and every one of his acts, should attain the proper end ([15], p. 120).

When any person, aspect, or part of a person is used "to serve away from his proper end", that use is immoral, perverse, and unnatural. It is perverse and unnatural because it is contrary to ends for which God designed nature. It is immoral because it is contrary to God's plan for humanity.

Masturbation, Thomas argues, is a paradigm case of the perverse, improper, unnatural abuse of an organ.

> Now, though the male semen is superfluous in regard to the preservation of the individual, it is nevertheless necessary in regard to the propagation of the species. Other superfluous things, such as excrement, are not at all necessary; hence their emission contributes to man's good. Now, this is not what is sought in the case of semen, but, rather, to emit it for the purpose of generation, to which purpose the sexual act is directed. But man's generative process would be frustrated unless it were followed by proper nutrition, because the offspring would not survive if proper nutrition were withheld. Therefore, the emission of semen ought to be so ordered that it will result in both the production of proper offspring and in the upbringing of this offspring.
>
> It is evident from this that every emission of semen, in such a way that generation can not follow, is contrary for the good for man. And if this is done deliberately, it must be a sin ([15], pp. 120–121).

Thus, for Thomas, since the end, or goal, of the sexual act is essentially procreative, any essentially nonreproductive sexual act – e.g., contraception or masturbation – is necessarily contrary to the ends of sexuality and thus perverse and unnatural. Moreover, as violations of God's ends in nature, such acts are not only unnatural and perverse, they are also sinful.

Sin is a concept that has been trivialized in the popular mind and ignored by most secular moral theorists. Yet the concept indicates a degree of moral transgression more extreme than the "wrong" of secular moral theory. For sins differ from other categories of moral wrongs because they are inexcusable except by divine intercession. While a non-sinful moral transgression can be excused, pardoned, or rectified by simple human action (e.g., by human forgiveness, contrition, pardon, or punishment) only God can pardon a sin. Consequently, by interpreting the sexually unnatural as the morally sinful, Thomas places extremely stringent moral sanctions on unnatural sexual acts.

The Sexual Romanticism of Immanuel Kant

The third of the sexual philosophers whose views on masturbation we will consider is Immanuel Kant (1724–1804). One of Kant's ambitions was to create a morality, founded in reason, that was entirely independent of presumptions about the nature, or even the existence, of God. While such an ethic necessarily eschewed concepts of sin and redemption, it incorporated, in suitably altered form, many of the concepts of the Christian moral theologians, particularly those of Augustine. Thus Kant's theory retained the Augustinian idea of the conflict between mind and body, except that, for Kant, the conflict was played out in terms of the tension between our rational will and our bodily inclinations. This conflict sets one of the major problems that Kantian moral theory addresses: the problem of determining the conditions under which moral human interactions are possible.

Kant's moral theories are difficult to present in summary fashion. Basically, he sees human interaction as morally problematic because it is all too frequently prompted by our inclination to use other people as mere objects – i.e., as tools to be put to use for our own purposes. Yet for Kant, the primary moral imperative is to treat persons not merely as objects, but as moral subjects – i.e., as autonomous rational agents, with purposes and projects of their own, who, as moral subjects, deserve the same respect that we would have others accord to ourselves.

Thus, while morality constrains our inclinations to use others as mere objects, our inclinations always tempt us to ignore them as moral agents and to treat them as objects of exploitation.

The struggle between morality and bodily inclination is especially acute in the context of sex.

> Amongst our inclinations there is one which is directed towards other human beings. They themselves, and not their work and services are its Objects of enjoyment . . . We refer to the sex impulse . . . when a person loves another purely from sexual desire . . . far from being concerned with any concern for the happiness of the loved one, the lover, in order to satisfy his desire and still his appetite, may even plunge the loved one into a state of misery. Sexual love makes of the loved person an Object of appetite; as soon as that appetite has been stilled, the person is cast aside as one casts away a lemon which has been sucked dry. . . . Taken by itself [sexual love] is a degradation of human nature; for as soon as a person becomes an Object of appetite for another, all motives of moral relationship cease to function, because, as an Object of appetite for another, a person becomes a thing, and can be used as such by everyone ([15], pp. 154–155).

Kant believed that sexual relationships are only morally acceptable in the context of a unity of will, i.e., in the context of a romantic love that culminates in marriage.

> In this way two persons become a unity of will. Whatever good or ill, joy or sorrow, befall either of them, the other will share in it. Thus sexuality leads to a union of human beings, and in that union alone its exercise is possible. This condition of sexuality, which is only fulfilled in marriage, is a moral condition ([15], p. 160).

Thus romantic love, by making two persons "a unity of will," requires each partner to respect the other's projects (i.e., ends or goals), thereby transforming sexual intercourse into a morally sustainable activity. The only true test of love, for Kant, is marriage. So he concludes that any form of sex, without marriage, "exposes mankind to the danger of equality with the beasts" ([15], p. 156).

Kant regarded masturbation as a fundamentally demeaning form of sex. It does more than expose someone to the danger of "equality with the beasts." Beasts are purely creatures of inclination – they have no possibility of autonomous rational agency. Humans, however, have the capacity to act as moral agents. Thus when masturbators, like Fred, choose to relinquish their autonomy, they choose to become like beasts. But because Fred had a choice, where beasts do not, he degrades himself in a way that no beast could. As Kant put the point, in "onanism," (i.e., the sin of Onan, "spilling seed upon the ground") a "man set[s] aside his person and degrades himself below the level of the animals" ([15], p. 162).

It is perhaps worth remarking that for Augustine, as for Kant, only other-directed love can reconcile sex with morality. The substantive difference in their analyses is that for Augustine the other is divine, while for Kant the other is human. If this difference is noted, however, their analyses are remarkably similar and lead to essentially the same assessment of masturbation. Masturbation is self-directed and loveless sex. The act thus lacks the other-directed love required to reconcile sex with morality and is therefore morally degrading. For Augustine (and for Aquinas) the act is more than degrading, it is also sinful. Sin, however, plays no conceptual role in Kant's secularized morality. So, although the Kantian condemnation of masturbation is structurally similar to Augustine's, there is one substantive difference between them – an Augustinian puritan will conceptualize masturbation as both self-degrading and sinful, the Kantian will condemn the act as self-degrading, but not as sinful. The difference between these two condemnations may seem, at first, to be trivial. As will become apparent when we reconsider the case of Fred's mother, however, this is one of those distinctions which makes a substantive difference.

Naturalism and Its Critics

Today, of course, masturbation tends to be regarded as neither sinful, nor wicked, nor even self-degrading. As one textbook author notes, "the general social acceptability of the practice is also clearly on the rise . . . The polite tolerance of earlier marriage manuals has also given way to unabashed endorsement by writers of some of the currently popular sex manuals [The Sensuous Man, The Sensuous Woman]" ([12], p. 284). One of the reasons underlying the endorsement of masturbation in contemporary sex manuals is that these manuals subscribe to an ideology that can be characterized, in the technical vocabulary of the moral philosopher, as "naturalist."

Naturalist moral theories are those in which the normal and/or the natural are identified with the morally permissible, the morally appropriate, or the ideal. In many versions of theory, the good and the normal/natural tend to be conceptualized as one and the same. For the Naturalist, the identification of the natural and the good is not an empirical matter, it is a conceptual truth. Thus "natural" and "good" become semantically intertwined – within the naturalist framework – so that it becomes unthinkable that natural could be anything other than good. (Natural foods, for example, are taken by those who might be thought of as dietary naturalists to be wholesome, not on the basis of scientific data, but on purely conceptual grounds – a point recognized in those health food advertisements that blatantly presume that readers will accept, without question, the equation of the unhealthy with the unnatural.)

The eighteenth-century philosopher David Hume [10] and the twentieth-century philosopher G. E. Moore [13] developed the classical critique of Naturalism. They argued that values are independent of any natural state of affairs. Hence the arguments typically adduced by naturalists must be fallacious because they are premised entirely on facts but nonetheless contain value-laden conclusions. *I.e*, these arguments proceed from premises that state what *is* the case and then fallaciously leap to conclusions about what *ought* to be the case. Moore baptized this inference the "Naturalist Fallacy." Moore would hold, for example, that the sex therapists commit the naturalistic fallacy when they conclude that adolescent masturbation is morally permissible *simply because*, in point of fact, adolescent boys *normally* masturbate. For Moore, the simple fact that something normally *is* the case can never provide sufficient grounds for the conclusion that this behavior *ought* to be the case.

To appreciate the force of Moore's critique, entertain for a moment the supposition that, in point of fact, adolescent boys normally raped adolescent girls, or that they normally killed their fathers, or that they normally boiled and then ate their grandmothers – in short, imagine that some clear case of morally reprehensible behavior were "normal." If, in spite of its normality, the behavior in question still appears to be reprehensible, then, as far as you are concerned, normality does not guarantee morality. If, for example, the discovery that any or all of these modes of behavior was normal would not suffice to convince you that it is morally permissible for boys to engage in these sorts of actions, then you are not a Naturalist. Most of us who have thought through the issue are not Naturalists. We know, for example, that adolescent boys are normally destructive of property; but no one seriously proposes that the normality of such destructiveness confers moral permissibility. Neither naturalness nor normality, in and of themselves, can make a state of affairs morally appropriate. The statistical normality of adolescent masturbation, therefore, has little bearing on the morality of the behavior.

Interpreting Nature: The Problems

One characteristic of Naturalistic ethics that vexes its critics is the presumed indubitability of a morality that appears vindicated by nature itself. T. H. Huxley, in a famous essay published almost a century ago, pointed out that the reason why nature appears to vindicate the values of Naturalists is that Naturalists conveniently project their own values onto the normative neutrality of nature. They thus discover in "nature" just the values that they project into it. To quote Huxley:

Viewed under the dry light of science, deer and wolf are alike admirable; and, if both were non-sentient automata, there would be nothing to qualify our admiration of the one upon the other. But the fact that the deer suffers, while the wolf inflicts suffering, engages our moral sympathies. We should call men like the deer innocent and good, men such as the wolf malignant and bad; we should call those who defended the deer and aided him to escape brave and compassionate, and those who aided the wolf in his bloody work base and cruel. Surely, if we transfer these judgments to nature outside the world of man at all we must do so impartially. In that case, the goodness of the right hand which helps the deer, and the wickedness of the left hand which eggs on the wolf, will neutralize one another: and the course of nature will appear to be neither moral nor immoral, but non-moral ([11], pp. 329–330).

One factor in the sex therapists' characterization of masturbation as normal is their projection of hedonistic values onto nature. Not everyone is content to cloak all forms of hedonism in the mantle of nature. Consider another case from the files of Dr. Cross ([1], pp. 461–462). This one concerns a thirty-three-year-old housewife, the mother of three children, who, during a rectal examination that she had requested, told the following tale.

Herb and I have a pretty good sex life, but lately we've gotten bored with the old routine and have been trying out new techniques. I found I liked it when he kissed or fingered my anus, so we decided to try intercourse there. But at first he couldn't get in, and the harder he tried the more it hurt. Finally he smeared butter on both of us, draped me over the back of a chair and drove in, but it was no fun for me. Then for a while he stayed quiet, and it began to feel nice and cozy. He hugged me tight and kissed my back and fingered my nipples and vulva. But as soon as he started to pump, it hurt again, and I was mighty glad when he finally finished and pulled out. He said it felt wonderful to him, much tighter than my vagina, but for me it was no fun at all. I lay awake for half an hour, feeling as if I'd been split apart, and I was still sore the next morning.

A nurse I know pretty well told me I was a fool to let myself be used that way for Herb's pleasure. She said anal intercourse is not only uncomfortable, but also messy and unsanitary. She offered to teach me some exercises which would make my vagina plenty tight enough and suggested that if Herb wasn't satisfied with that, there must be something wrong with him.

Herb says it takes time to get used to this approach, and we should keep trying, but I'm not about to go through that experience again. I wondered if perhaps I had something wrong down there, so I came to see you.

Dr. Cross also reports that the rectal examination revealed no hemorrhoids, fissures, fistulas or other abnormalities – and that the sphincter tone was good.

How would a sexual naturalist respond to this case? Clinicians who used Godow's textbook on human sexuality might recall the following discussion.

Since the anus is rich in nerve endings and is contiguous with the sexually sensitive genitalia in both sexes, it is not surprising that the anal area may derive erotic significance. Some people enjoy the tactile stimulation of the anal area as a foreplay technique, and some couples, both heterosexual and homosexual, engage in anal intercourse (penile penetration of the anus) as one variation of their repertoire of sexual activity. . . .

Recent studies suggest that increasing numbers of married couples are exploring anal intercourse as a variation to penile-vaginal intercourse. . . .[a]bout one seventh of the 35-to-44 age group and one fourth of the under-35 age group had done so. . . .In the *Redbook* study 43% of the married women said that they had tried anal intercourse at least once. . . . Of those who tried anal intercourse, 41% described the experience as somewhat or very enjoyable, 49% characterized it as unpleasant or repulsive, and the remainder reported having neutral feelings. . . .[i]t appears as though heterosexual couples are experimenting with anal intercourse more often. . . .

Anal intercourse is sometimes referred to as sodomy . . .and is against the law in the majority of states. Fortunately these so-called sodomy laws are seldom enforced. . . ([9], pp. 143–144).

Herb would appear to be among that 41% of experimenters who found anal intercourse enjoyable. His wife is among the 49% of the experimenters who found the experience unpleasant. What implications do these data have for sexual naturalists? Do they tell us that anal intercourse is right for 41% of all people but wrong for 49%? Or should one ignore the fact that about half of all those who engage in the practice appear to enjoy it, and condemn the practice?

Consider what one eighteenth-century proponent of Sexual Naturalism, the Marquis de Sade, had to say about these matters. After citing the Naturalist's creed –

Nature wills it. For a bridle have nothing but your inclinations, for laws only your desires, for morality Nature's alone . . . ([15], p. 301).

De Sade addresses the issue of anal intercourse, or sodomy:

> But sodomy, that alleged crime which will draw the fire of heaven upon cities addicted to it, is sodomy not a monstrous deviation whose punishment could not be severe enough? . . . .Nature, who places such slight importance upon the essence that flows in our loins can scarcely be vexed by our choice when we are pleased to vent it into this or that avenue. . . .What single crime can exist here? For no one will wish to maintain that all parts of the body do not resemble each other, that there are some which are pure, and others defiled; but as it is unthinkable such nonsense be advanced seriously, the only possible crime should consist in the waste of semen. Well, is it likely that this semen is so precious to Nature that its loss is necessarily criminal? Were that so, would she every day institute those losses? and is it not to authorize them to permit them in dreams, to permit them in the act of taking one's pleasure with a pregnant woman? Is it possible to imagine Nature having allowed us the possibility of committing a crime that would outrage her? . . . It is unheard of – into what an abyss of folly one is hurled when, in reasoning, one abandons the aid of reason's torch! Let us abide in our unshaken assurance that it is as easy to enjoy a woman in one manner as in another, that it makes absolutely no difference whether one enjoys a girl or a boy, and as soon as it is clearly understood that no inclination or tastes can exist in us save the ones we have from Nature, that she is too wise and too consistent to have given us any which could ever offend her ([15], pp. 304–305).

De Sade argues that a true Naturalist – i.e., someone who takes for his morality Nature's alone – acting by the pure light of "reason's torch," has no reason not to "enjoy a woman in one manner [anally] as in another [vaginally]." Naturalists will enjoy anyone, via any orifice, that he or she is inclined to use. He points out that since all parts of the body resemble each other, none can be "purer" than the other – all are equally nature's products. Moreover, if a man, e.g., Herb, has an inclination to enjoy a woman through her anus, that inclination must come from nature, for "no inclination or tastes can exist in us save the ones we have from nature." As for the argument that anal intercourse is uncomfortable, messy, and unsanitary, De Sade would, no doubt, have pointed out that the same could be said for intercourse with a virgin or, for that matter, about childbirth – yet these are not considered unnatural on that account.

Perhaps the argument will be made that nature intended all organs and orifices to be unifunctional: the mouth for eating, the vagina for intercourse, and the anus for defecating. A moment's reflection on the dual functions of the mouth (talking-eating), as well as the penis and the vagina, will establish that nature permits organic multifunctionality.

The primary argument against anal intercourse, as De Sade points out, is that this mode of intercourse cannot be what nature intended *because* the act is essentially non-procreative. But how does one know that nature intended no non-procreative sexual acts? Perhaps nature intended some sexual activities to be non-procreative? Perhaps anal intercourse is a natural form of birth control? Alternatively, if anal intercourse is unnatural because nature intended intercourse to be procreative, then Naturalists must agree with Dr. Manson (and A. J. Cronin) that contraception, too, is unnatural – and so too is masturbatory sex, which is equally non-procreative.

It is difficult to condemn Herb for engaging in non-procreative sex without also condemning Fred. From a certain perspective, they were both engaged in essentially the same activity; only Fred used a book, and, perhaps a pillowcase or a blanket, while Herb used a woman's anus. To argue that the former was unnatural while the latter is not would be to contend that a book and a blanket are more natural than a woman and her anus. Anal, contraceptive, and masturbatory sex are all equally non-procreative; hence, if procreativity is the standard of naturalness, one must consider all or none, equally, to be either natural or unnatural. One cannot distinguish between them without falling into inconsistency.

Nature knows no inconsistencies. The inconsistencies we find there are products of the interpretations we foist upon it when we reify our values as natural laws. In truth, however, neither nature, nor anatomy, nor any study of the prevalence of sexual practices can tell us whether it is proper to masturbate or to use contraceptives or to engage in anal intercourse. The answers to these questions, insofar as they have answers, can only be discovered in philosophical reflection.

The Case of Fred's Mother Reconsidered

What are clinicians to do when a patient provokes a question that challenges the values embedded in modern medicine? One option – the option exercised by Dr. Manson and by the nurse who informed Herb's wife that she was "a fool" to let herself "be used in that way for Herb's pleasure" – is to treat this as an occasion for ordinary moral discourse, *i.e.*, an occasion for judging another's moral character ("fool," "dirty little man of God"). The privileged nature of the clinician-patient relationship, however, militates against the intrusion of ordinary morality. The clinic has its own morality, a morality where the good is defined as the "healthy" and badness becomes "pathology," a morality that rests above all else on the presumption of a shared normative framework. Patients come to the clinic anticipating this framework. And, as the Manson-Parry case illustrates so well, when moral dissension violates these expectations, the clinic loses its moral prerequisites and the clinician-patient relationship collapses.

The conventional wisdom would have clinicians choose a second option – an option in which clinicians neither impose their own values on their patients, nor accept their patient's values, nor enter into ordinary moral discourse. Conventional wisdom would have clinicians avoid valuational issues altogether and simply stick to facts. Such a posture is problematic because it ignores the normative saturation of medicine. Clinicians can no more avoid values than they can eschew the concept of pathology. A truly scientific clinician could not help Fred's mother or Herb's wife or even Reverend Edwal. Questions about who should be treated, Fred or his mother, Herb or his wife, the question of whether or not to give Edwal information and assistance about contraception, are not scientific questions. Attempts to be "scientific" about these matters merely allow clinicians to slide inadvertently into naturalism.

Only one alternative remains. When faced with normative dissent, clinicians should neither flee to science nor fall into ordinary moral discourse, instead they should face normative questions forthrightly and philosophically. Ours is an era of sexual revolution. The biological connections between intercourse, conception, birth, and nurturance have been severed; the cultural institutions surrounding sex, gender roles, courtship and romance, marriage, and family have changed. Ideas and practices are all uncertain and so, not unnaturally, people are confused. In their confusion, they sometimes turn to clinicians. They frequently seek neither cure nor conversion, but only a philosophical forum in which they can safely assess the implications of the sexual revolution on their personal lives.

Clinicians should respond to the quest of such patients by creating a forum for the analysis and articulation of values – a forum for the exploration of various conceptions of anal intercourse and masturbation. For example, the nurse who condemned Herb's wife for allowing herself to be treated as a sex object might instead have gotten her patient to articulate her own views on sex; then she might have drawn out the implications of those views as they applied both to anal intercourse and to the asymmetry between Herb's pleasure and her patient's pain. Such an analysis would have been invaluable to the patient by giving her a sanctuary from pre-judgment (prejudice) in which she could freely explore her own thoughts and develop her own valuation of anal intercourse.

Fred's mother also needed a sheltered forum in which she could explore her valuational disagreements with her husband and her ambivalence towards the general societal acceptance of masturbation. She needed to reflect on the reasons for her feelings of shame; she needed to decide whether she believed that masturbation was sinful; she needed to reflect on adolescence, that period in our lives when we engage in so many activities we later rue; she needed to think through the encounter with our animal nature that is central to mature emergence from adolescence. She went to a clinician, but she sought a philosopher. And, in the circumstances, her clinician could have served her best by playing the role of sexual philosopher.

Union College,
Schenectady, New York, U.S.A.

BIBLIOGRAPHY

1. Baker, R., and F. Elliston: 1984, *Philosophy and Sex*, rev. ed., Prometheus Books, Buffalo.
2. Barker-Benfield, G.: 1976, *The Horrors of the Half-Known Life: Male Attitudes Towards Women and Sexuality in Nineteenth Century America*, Harper and Row, New York.
3. Boswell, J.: 1980, *Christianity, Social Tolerance, and Homosexuality*, University of Chicago Press, Chicago.
4. Caplan, A., *et al.* (eds.): 1981, *Concepts of Health and Disease: Interdisciplinary Perspectives*, Addison-Wesley, Reading.
5. Clendening, L.: 1960, *Source Book of Medical History*, Dover Publications, New York.
6. Cronin, A. J.: 1937, 1965, *The Citadel*, Little Brown & Co., New York.
7. Dover, K.: 1978, *Greek Homosexuality*, Harvard University Press, Cambridge.
8. Foucault, M.: 1978, *The History of Sexuality*, Vol. 1, tr. R. Hurley, Pantheon Books, New York.
9. Godow, A.: 1982, *Human Sexuality*, C. V. Mosby, St. Louis.
10. Hume, D.: *A Treatise On Human Nature*, ed. L. Selby Bigge, Oxford University Press, Oxford.
11. Huxley, T.: 'The Struggle for Existence in Human Society', *The Nineteenth Century* 23 (Feb. 1888), 161–180, reprinted in P. Kropotkin: 1976, *Mutual Aid*, Appendix B, Porter Sargent, Boston, pp. 329–330.
12. Katchadourian, H.: 1975, *Fundamentals of Human Sexuality*, 2nd ed., Holt, Rinehart and Winston, New York.

13. Moore, G.: 1903, 1962, *Principia Ethica*, Cambridge University Press, Cambridge.
14. McCary, J.: 1979, *Human Sexuality*, 2nd ed., Van Nostrand, New York.
15. Verene, D. (ed.): 1972, *Sexual Love and Western Morality: A Philosophical Anthology*, Harper and Row, New York.

FREDERICK SUPPE

THE DIAGNOSTIC AND STATISTICAL MANUAL OF THE AMERICAN PSYCHIATRIC ASSOCIATION: CLASSIFYING SEXUAL DISORDERS[1]

> Health consists in having the same diseases as one's neighbors.
>
> Quentin Crisp [13]

The *Diagnostic and Statistical Manual* (DSM) ([1], [2]) of the American Psychiatric Association (APA)[2] and the World Health Organization (WHO)'s *International Classification of Diseases* (ICD) ([40], [41]) list as mental disorders (DSM) or diseases (ICD): homosexuality; ego-dystonic homosexuality; fetishism; pedophilia; transvestitism; exhibitionism, voyeurism; sexual sadism; sexual masochism; zoophilia; transsexualism; gender identity disorder of childhood; various psychosexual dysfunctions such as inhibited sexual desire, excitement, and orgasm; and disorders of psychosexual identity including transsexualism and feminism in boys ([2], pp. 380–381, 429–431). Judged by currently fashionable lists of sexual aberrations (e.g., [21], [22]) and issues of *The Fetish Times*, the lists are curiously incomplete; although caffeine intoxication is included in DSM-III, rape[3] and incest are not. These omissions are especially bewildering since DSM-III includes such more esoteric sexual disorders (under "Atypical Paraphilia" (320.90)) as coprophilia, frotteurism, klismaphilia, mysophilia, necrophilia, telephone scatologica, and urophilia (respectively feces, rubbing, enema, filth, corpse, obscene-phone-call, and urinary perversions).

Prior to the 1948 and 1953 Kinsey reports ([17], [18]), the social mores in our society were such that there would have been little difficulty getting near unanimity of opinion that these sexual practices catalogued in DSM-III and ICD-9 were the marks of mentally ill persons. That secure feeling of sexual normality was blown by the Kinsey and later

Earl E. Shelp (ed.), Sexuality and Medicine, Vol. II, 111–135.

reports, which reported hugely unsuspected incidences of homosexuality and bestiality, and growing suspicions that the more esoteric acts mentioned above had significant advocates.[4]

The growing post-Kinsey-Reports sexual liberation has resulted in champions of the previously illicit sexual behaviors joining in various sexual liberation movements, including such esoteric ones as the masochists' liberation movement (e.g., the *Til Eulenspiel* society in New York) and the pederasts' liberation movement (e.g., the North American Man/Boy Love Association; see [23], [39]), but also the more mainstream and quite successful Gay Liberation movement. Gay Liberation has not only been a major influence in the decriminalization of homosexual behavior in dozens of states, but it also played a major role in the removal of homosexuality from the APA's catalogue of mental disorders (DSM) ([3], [34]), and it serves as an impetus to other sexual liberation movements. Those debates surrounding the reclassification of homosexuality as a mental disorder in DSM-II, and its aftermath in the drawing up of DSM-III (which does not include homosexuality but includes ego-dystonic homosexuality), raise not only questions about the psychiatric evaluation of homosexuality, but also the psychiatric evaluation (as reflected in DSM-III) of sexual disorders in general. Specifically, there is the issue of the extent to which such psychiatric evaluations are just pseudo-scientific/medical masquerades of prevailing or reactionarily-conservative social mores. As a means to exploring this issue we will first look in some detail at the controversial APA decision to exclude homosexuality from DSM-II and III, while adding "ego-dystonic homosexuality," and then turn to a consideration of the objectivity of including other sexual paraphilias as mental disorders in DSM-III.

EGO-DYSTONIC HOMOSEXUALITY AND THE CONCEPT OF MENTAL DISORDER

In large part as a result of Gay Liberation lobbying, which stressed research (by Evelyn Hooker and others) challenging the prevailing psychiatric orthodoxy that psychopathology was an inevitable concomitant of homosexuality, in 1973 the American Psychiatric Association removed homosexuality from its catalog of mental disorders (DSM-II). This decision, challenged by a group of psychiatrists headed by Irving Bieber and Charles Socarides, forced a referendum vote of the APA membership on the issue, Gay Activists became involved in attempts to defeat the referendum, and the removal stood: Homosexuality no longer was a mental disorder, although "Sexual Orientation Disturbance" (homosexuals who are "disturbed by, in conflict with, or wish to change their sexual orientation") was added as a mental disorder [31].[5] Much of the notorious dispute having been carried out in or reported by the press, the episode was widely viewed as a breakdown of scientific objectivity where power politics and vested interests within and outside of psychiatry replaced reasoned debate and the canons of scientific evidence.

In 1974 Robert Spitzer was appointed Chair of the APA Task Force on Nomenclature and Statistics, which was charged with preparing a new edition of the *Diagnostic and Statistical Manual* – DSM-III. One of the more positive effects of the APA controversy over declassifying homosexuality was that it prompted Spitzer and his committee to examine carefully and rethink critically the very notion of a mental disorder.

That such an examination was in order was made patently obvious during the APA controversy by the fact that while many protagonists frequently used such terms as "disease," "illness," "disorder," "sickness," "pathological," "dysfunction," "disturbance," etc. uncritically as interchangeable synonyms, others tried to draw (largely unarticulated) distinctions between these. For example, whereas in a single article [27] Charles Socarides variously refers to homosexuality as a "form of mental illness," a "medical problem," "a dread dysfunction," and "illness diagnosed as perversion," "a form of psychiatric or emotional illness," an "indication of a pathological condition" and concludes "there is no obligatory homosexual who can be considered to be healthy" – others such as Irving Bieber maintained that homosexuality should be classified as a mental disorder but (retrospectively, at least) deny ever "having written or said that all homosexuals are disturbed – much less that all homosexuals are seriously disturbed" [5]. While homosexuality is a sexual deviation and a manifestation of psychopathology for Bieber, it is not a mental illness since "the term 'mental illness' connotes psychosis. . . . I know of no psychiatrist who believes homosexuality is a psychosis." Non-psychotic conditions such as sexual deviation, including homosexuality, are "not . . . illnesses, but . . . mental disorders" [6], pp. 18, 20, 22). Against Gay Activist charges that retention of homosexuality in DSM as a mental disorder stigmatizes homosexuals as "sickies," Bieber's rejoinder is that homosexuality is neither a sickness nor an illness and that homosexuals are not inevitably disturbed; they just possess a psychopathological mental disorder characterized by behavior "based on irrational and unrealistic fears . . . of lethal attack by other men should heterosexuality be attempted or contemplated" (*ibid.*, pp. 11–12). One gets the distinct impression that obfuscatory semantic trickery may be going on.

Such an impression is reinforced when one considers that DSM-II was specifically designed to integrate with WHO's *International-Classification of Diseases* (ICD-8) under Section V, "mental disorders"; *prima facie*, then, in DSM-II mental disorders are viewed as a species of diseases. Further, if one consults standard medical dictionaries such as Taber's [36], one finds *mental disease* defined as "a disorder of the mind or intellect," *mental illness* as "any disorder which affects the mind or behavior," and *pathologic* as "diseased." Nowhere, however, does one find the terms *disorder* or *mental disorder* defined, although they are used in other definitions. In standard medical parlance, then, it would appear that mental disorders and psychopathologies merely are mental diseases or illnesses – and that Bieber's usages are just idiosyncratic.

Philosophers are acutely aware that dictionary definitions rarely settle issues of conceptual analysis; and so it would be a mistake to think the issue is resolved. Further, there is a rich philosophical literature concerned with analyzing the concepts of *disease* and *illness* – and there is mounting reason to think the two are distinct but related concepts [15]. So much for common medical parlance settling the issue. Moreover, if one goes through the DSM-III classification of mental disorders, one frequently encounters disorders which, to my mind, are not plausibly construed as diseases or illnesses, although they plausibly are construed as short-term disorders. Examples include alcohol intoxication (303.00) as opposed to alcoholism (303.9), cocaine intoxication (305.60); and cannibus intoxication (305.20). And while tobacco dependence (305.1) involves addiction, I am uncomfortable calling it a disease or an illness. Thus, e.g. in DSM-III not all mental disorders are mental illnesses or mental diseases – although some are.

What, then, is a mental disorder? The problem becomes more acute when I try to fathom why DSM-III classifies Developmental Reading Disorder (315.00) and Atypical Specific Developmental Disorder (315.90) as mental disorders but classifies Academic Problem (V62.30) as a "Condition not Attributable To a Mental Disorder" that is a focus of attention or treatment. The diagnostic criteria are as follows:

Developmental Reading Disorder: Performance on standardized, individually administered tests of reading skill is significantly below the expected level, given the individual's schooling, chronological age, and mental age (as determined by an individually administered IQ test). In addition, in school the child's performance on tasks requiring reading skills is significantly below his or her intellectual capacity (p. 94).

Academic Problem: This category can be used when a focus of attention or treatment is an academic problem that is apparently not due to a mental disorder. An example is a pattern of failing grades or of significant underachievement in an individual with adequate intellectual capacity, in the absence of a Specific Developmental Problem Disorder or any other mental disorder to account for the problem (p. 332).

Atypical Specific Developmental Disorder: This is a residual category for use when there is a Specific Developmental Disorder not covered by any of the previous specific categories (p. 99).

Exactly what a mental disorder is becomes increasingly unclear.

Spitzer and his committee were acutely aware how problematic the concept of mental disorder is, and went to considerable effort to try to produce a clear analysis of the concept. Spitzer has written:

When I first was given the job of considering the claims . . . that homosexuality should not be regarded as a mental disorder, I was confronted with the absence of any generally accepted definition of mental disorder. I therefore reviewed the characteristics of the various mental disorders and concluded that, with the exception of homosexuality and perhaps some of the other "sexual deviations," they all regularly cause subjective distress or were associated with generalized impairment in social effectiveness or functioning. It became clear to me that the *consequences* of a condition, and not its aetiology, determined whether or not the condition should be considered a disorder ([29], p. 5).

Spitzer and others went on to develop a very complicated definition of mental disorder replete with specific criteria, but ultimately became convinced it was unsatisfactory and that "no precise definition of disorder (physical or mental) was possible or even useful" (ibid., p. 7). However, DSM-III does present "concepts" that have influenced decisions to include certain conditions in DSM-III while excluding others – *viz*:

In DSM-III each of the mental disorders is conceptualized as a clinically significant behavioral or psychological syndrome or pattern that occurs in an individual and that is typically associated with either a painful symptom (distress) or impairment in one or more important areas of functioning (disability). In addition, there is an inference that there is a behavioral, psychological, or biological dysfunction and that the disturbance is not only in the relationship between the individual and society. (When the disturbance is *limited* to a conflict between an individual and society, this may represent social deviance, which may or may not be commendable, but is not by itself a mental disorder.) (p. 6)

The repercussions on this criterion of the earlier controversy over declassifying homosexuality as a mental disorder are overt, and it does provide fairly explicit criteria for excluding homosexuality, left-handedness, and various forms of uncouth behavior from the list of mental disorders. But it obviously is not wholly adequate. For it does nothing to answer my questions, raised above, why Developmental Reading Disorder and Atypical Specific Developmental Disorder are mental disorders, but Academic Problem is not. And, as we will see below, its employment with regard to the inclusion of the sexual paraphilias and Ego-dystonic Homosexuality is rather problematic. Before turning to such issues, some consequences of this criterion need to be explored.

As Spitzer has noted, no precise definitions are given for key notions such as "impairment in one or more important areas of functioning" and "behavioral, psychologic, or biologic dysfunction" ([29], p. 8) and that "It should be understood that there is always a value judgement in deciding that a particular area of functioning is 'important'" (*ibid.*, p. 11). In short, when subjective distress is absent, whether a condition is a mental disorder ultimately is a value judgement and not a purely factual matter. Those such as Boorse [8] who hold out for non-normative conceptions of disease and illness surely will find this wholly unacceptable. And given the extent to which judgements of what is mentally healthy by clinicians so very strongly reflect conformity to sex-role stereotypes [9], such a criterion of mental disorder would seem to license the use of psychiatry as a means for enforcing social, political, and ideological conformity in ways that are commonplace in the Soviet Union, and which Gay Activists perceived psychiatry as doing so long as homosexuality remained on the list of mental disorders. To be sure, Spitzer does not approve of such psychiatric totalitarianism, and the criteria do include specific caveats designed to block such misuses. But the effort seems to me doomed – for how does one distinguish impaired psychological functioning from "mere conflicts between an individual and society"? Spitzer seems to rely on consensus among psychiatrists and what he terms "the concept of inherent disadvantage" ([21], pp. 11

ff.), but I do not see that these really come to grips with the issue. For it would appear that prior to Gay Liberation, when there was consensus that homosexuality was dysfunctional with inherent disadvantage, it *was* a mental disorder; but *now* that the educative and activist activities of Gay Liberation have raised doubts in so many therapists' minds and destroyed professional consensus, it no longer *is* a dysfunctional condition but rather is merely a conflict between society and homosexual individuals. Spitzer himself has qualms similar to my own (*ibid.*, pp. 20–22).

Do such potential misuses of psychiatric diagnosis show that a normative, value-laden notion of mental disorder *ipso facto* is defective? No, for *any* notion of mental disorder, illness, or disease *must* be normative. Freudian psychology, as well as many other psychoanalytic theories, specifies a "normal" pattern of psychological development that stresses typical patterns of interaction with one's environment, as well as various patterns of "abnormal" psychological development or influences augmented by accounts of how to manipulate the environment so as to affect one's psychological state. Psychotherapeutic theories such as Rogers', which eschew "normal patterns of development," otherwise are similar in that they view an individual as having a particular psychological/behavioral state and being in an environmental state, and the theory describes how subsequent psychological states are a function of prior psychological and environmental states. In general, psychotherapeutic theories construe individuals as teleological systems with laws of quasi-succession.[6] Teleological theories specify certain of the (e.g.) psychological states as *goal states*. Goal states can be naturalistic or normative. Naturalistic goal states are those states (e.g., psychological states) that a system will *tend towards*, given a *stable* environment. Normative goal states generally constitute some *other* set of goal states than the naturalistic ones, which are viewed as desirable on some criterion. In psychoanalytic theories the naturalistic goal states include various pathological, dysfunctional states, which, according to the theory, result from stable but undesirable environmental influences during developmental years (Bieber's account of homosexual aetiology [6], [7] is a perfect example). But psychoanalysis is concerned with effecting mental health, which is to say that it has a *normative* set of goal states that excludes pathological states. Any plausible psychotherapeutic theory thus must construe mental health or the absence of mental disorder as possession of normative psychological/behavioral goal states.

Thus DSM-III's criteria for determining what are mental disorders is not inadequate merely in virtue of making it a normative value-laden issue; for it ultimately must be. Rather, it must be assessed on grounds of its values, on its normative adequacy. Precisely these issues were raised in the development of DSM-III in the context of debates over the tenability of "Sexual Orientation Disturbances" as a mental disorder. Ultimately, DSM-III did include such a notion reworded as "Ego-dystonic Homosexuality" (302.00), which has the following diagnostic criteria:

A. The individual complains that heterosexual arousal is persistently absent or weak and significantly interferes with initiating or maintaining wanted heterosexual relationships.

B. There is a sustained pattern of homosexual arousal that the individual explicitly states has been unwanted and a persistent source of distress (p. 282).

Many experts (e.g., Paul Gebhard and John Money) argued against the inclusion of ego-dystonic homosexuality in DSM-III, suggesting that if this were a legitimate category, it is arbitrary to include it but exclude ego-dystonic heterosexuality, "distress over adulterous impulses," "ego-dystonic masturbation," and so on; while others thought it appropriately should be included under extant categories such as "Psychosexual Disorders not Elsewhere Classified" or various anxiety categories. Spitzer has various responses to this line of objection ([29], pp. 15–18), of which only the claim that there is no recorded instance of ego-dystonic heterosexuality initially seemed convincing. Since then I have encountered a clear case of ego-dystonic heterosexuality in my counselling practice.[7]

But Spitzer's responses do not address themselves to the more general problems concerning "ego-dystonic" categories. If ego-dystonic homosexuality, then why not also ego-dystonic psoriasis (better known as "the heartbreak of psoriasis"), ego-dystonic hemorrhoiditis (promoted by Preparation-H), ego-dystonic acnitis (promoted by Pat Boone, Clearasil, *et al.*), and ego-dystonic acomia (promoted by hair weaving, hair transplants, and wig firms). In short, the range of symptoms that can be recognized by an individual as unacceptable and undesirable and is experienced as alien – that is, can be *ego-dystonic* – boggles the imagination, and a decision to include just *one* ego-dystonic condition as a mental disorder seems *highly arbitrary* and unwarranted.

To return to our earlier discussion of the difficulties in distinguishing mental illnesses or diseases from short-term disorders, it is important to consider the phenomenon of "coming out" to oneself as homosexual ([30], pp. 76–78, 83–86; [12], [14]). While some homosexuals report knowing they were homosexual as long as they have been alive, another common pattern is for an individual to believe him/herself to be heterosexual, yet encounter frequent episodes of interest in same-sex bodies and physiques/figures, erotically fantasize about members of the same sex, and even engage in homosexual activity – all the while engaging in a variety of defense mechanisms designed to convince themselves they are really heterosexual. At times the individual may precisely fit the ego-dystonic homosexuality criteria quoted above. Assuming the individual truly has a homosexual orientation, a standard scenario is for the individual eventually to confront his/her homosexuality, come to accept it, and begin working on integrating him/herself into the homosexual subculture and develop a happy, satisfying life style. Doing so effectively eliminates the ego-dystonia of one's homosexuality. The point is that in this frequent scenario difficulty in accepting one's homosexuality leads to temporary subjective distress and possible dysfunctionality, but constitutes neither a long-term disorder or disease. Whether it is a disorder any more worthy of inclusion in DSM-III than the omitted ego-dystonic acnitis (which sometimes is of longer duration) is unclear. It is worth noting here that large numbers of psychotherapists are of the opinion that when confronted by a case of ego-dystonic homosexuality the preferred strategy is to help the client accept his/her homosexuality and integrate it into a productive and rewarding life-style.[8]

Short of highly cynical charges that the inclusion of ego-dystonic homosexuality, but not other ego-dystonias, is purely a matter of fee-protectionism by psychiatrists specializing in the "cure" of homosexuality, we are drawn back to the anomalies of lumping short-term problems with mental illness or disease and the exclusion of significant ego-dystonias ranging from the often serious crises in the aftermath of rape and incest to distress over baldness, acne, etc., but the inclusion of ego-dystonic homosexuality as the *only* ego-dystonia in DSM-III. All of this leaves unanswered the suspicion, mentioned above, that the inclusion of ego-dystonic homosexuality is a codification of the social mores of the sexually more conservative. Such fears are reinforced by a consideration of DSM-III's treatment of the so-called "sexual paraphilias" which are even more unconventional and more in conflict with prevailing sexual mores than is homosexuality. In particular, the question arises whether the other sexual paraphilias are any less a social deviance, which is limited to a conflict between an individual and society, than is homosexuality – which was so judged in its deletion from DSM. If so, should not other sexual paraphilias be deleted from DSM? And if so, is their replacement by ego-dystonic versions justified? We now turn to these issues.

SEXUAL PARAPHILIAS

One of the main groups of psychosexual disorders in DSM-III is the paraphilias. "The paraphilias are characterized by arousal in response to sexual objects or situations that are not part of *normative* arousal-activity patterns and that in varying degree *may interfere* with the capacity for reciprocal affectionate sexual activity" (p. 261; italics added). Thus, at the very heart of DSM-III's thinking about the paraphilias is the idea that deviation from conventional sexual activity (intercourse possibly preceded by foreplay) as an expression of affection and love is unhealthy.[9] The paraphilias listed in DSM-III are: Fetishism (302.81), Transvestism (302.30), Zoophilia (302.10), Pedophilia (302.20), Exhibitionism (302.40), Voyeurism (302.82), Sexual Masochism (302.83), Sexual Sadism (302.84), and Atypical Paraphilia (302.90), which "is a residual category for individuals with Paraphilias that cannot be classified in any other categories. Such conditions include: Coprophilia (feces), Frotteurism (rubbing); Klismaphilia (enema); Mysophilia (filth); Necrophilia (corpse); Telephone Scatologica (lewdness); and Urophilia (urine)" (p. 275).

The term *paraphilia* literally means "craving for the abnormal," but in sexology it has acquired a rather more specific meaning where it signifies the *necessity* or psychological dependency of such unusual objects or acts or fantasies thereof for arousal and orgasm, as opposed to the mere ability to respond erotically to such unusual objects or acts ([20], p. 220; [22], pp. 342–343). DSM-III's characterization of paraphilia is in conformity with this standard usage:

> unusual or bizarre imagery or acts are *necessary* for sexual excitement. Such imagery or acts tend to be insistently and involuntarily repetitive and generally involve either (1) preference for use of a non-human object for sexual arousal, (2) repetitive sexual activity with humans involving real or simulated suffering or humiliation, or (3) repetitive sexual activity with nonconsenting partners (p. 266; italics added).

But when DSM-III gets down to the specific diagnostic criteria for paraphilias such as fetishism (302.81), zoophilia (302.10), pedophilia (302.20), voyeurism (302.82), and certain forms of sexual masochism (302.83) and sexual sadism (302.84), the requirements are loosened so that the use of such sources of excitement only need be a "repeatedly preferred or exclusive method of achieving sexual excitement" (pp. 269, 270, 271–272, 273). For transvestitism (302.30), with respect to sexual excitement all that is required is "use of cross-dressing for the purposes of sexual excitement, *at least initially* in the course of the disorder" (p. 270; italics added). In the case of exhibitionism (302.40) and certain forms of sexual sadism (302.84), *repetitive* occurrences of the acts are sufficient (pp. 272, 275). In certain forms of sexual masochism (302.83) and sexual sadism (302.84), a *single* episode is sufficient (pp. 274–275). Further, while DSM-III's definition of a paraphilia puts actual activity and fantasy or imagery on a par, the diagnostic categories for fetishism (302.81), exhibitionism (302.40), voyeurism (302.82), and sexual masochism (302.83) require that the acts actually be carried out, and in most cases explicitly disallow fantasy alone from qualifying. In the case of sexual sadism (302.84), the actions must be actually carried out or else simulated. Only in zoophilia (302.10) and pedophilia are overt activity and fantasy (imagery) put on a par as the definition of paraphilia requires. Thus we see that the actual diagnostic requirements for specific paraphilias imposed by DSM-III generally do not meet DSM-III's own requirements for being a paraphilia.

When I brought some of these disparities to Robert Spitzer's attention during an extended discussion of the paraphilias, his response was to eliminate the disparities by changing the definition of "paraphilia." To do so would be unwise for several reasons. First, it does violence to the well-established standard clinical and research usage noted above. Second, the disparities between the definition of "paraphilia" and the actual clinical diagnoses DSM-III provides are rooted in inadequate understanding of human sexuality and the various "kinky" activities DSM-III is concerned with. Thus, third, accepting Spitzer's proposal would inhibit coming to a better understanding of these sexual behaviors and their clinical implications. I now turn to the substantiation of my second reason (hence of the third by implication).

PATTERNS OF SEXUAL RESPONSE

An individual's *sexual identity* has the following distinct components: *biological sex* (the determination whether one is male or female at birth); *gender identity* (one's basic conviction of being male or female); *social sex role* (extent of conformity to physical and psychological characteristics culturally associated with males and females); and *sexual orientation*, which includes *sexual behavior* (patterns of erotic bodily contact with others); patterns of *interpersonal affection* (associations involving various degrees of trust such as with friends, lovers and marital partners); *erotic fantasy structure* (sexually arousing patterns of mental images of one or more persons engaged in physical sexual activity or in affectional relationships), *arousal cue-response patterns* (which sensory cues stimulate or inhibit erotic arousal), and *sexual self-concept* (one's own conception or labeling of oneself as a sexual being [24], [31], [32], [33], [38]). These components may or may not be in conformity with each other. E.g., a person may be biologically male but believe himself to be really female, or one's social sex role may be sharply at odds with one's gender. Sexual orientation usually is labeled heterosexual, homosexual, or bisexual, but in fact these labels also can be assigned to its various components, which may not be in accord with each other. Thus, one can be homosexual in some aspects of his sexual orientation and heterosexual or bisexual in others.

Since we are concerned with the paraphilias here, it will be useful to expand our notions of sexual orientation to include more than just homosexuality, heterosexuality, and bisexuality. One's sexual behavior can be directed towards any of the paraphilias, as can one's fantasy structure or arousal cue-response pattern. And while there is less scope for paraphiliac patterns of interpersonal affection, I have counselled one individual who only could trust or love dogs with whom he was sexually active, and in his case I have no qualms labeling his interpersonal affection pattern as zoophiliac (see also below). Thus with respect to various components of sexual orientation, an individual may be heterosexual, homosexual, bisexual, fetishistic, transvestic, zoophiliac, pedophiliac, exhibitionistic, voyeuristic, masochistic, sadistic, etc. It is important to note that these are not exclusive categories. For example, a female who can only get aroused by being dressed in leather and put in bondage while fellating a male with a dog performing cunnilingus would be a heterosexual zoophiliac fetishistic masochist.

When a significant portion of the cues in one's arousal response pattern are highly correlated with features of the opposite sex, one is said to be heterosexual; when correlated with the same sex, one is said to be homosexual; and when either gender-neutral or else one develops separate heterosexual and homosexual cue-response patterns without contrary inhibitory cues, one is bisexual. But not all such cues are so gender-correlated. It is commonplace for persons to have cues that are inanimate or situational. And to the extent that one does, one can be said to be fetishistic, masochistic, sadistic, etc., in cue-response pattern.

Once we realize the diversity of animate, inanimate, and situational cues that routinely are found in individual cue-response patterns, for purposes of classification it is important to give consideration to how they fit into the overall response pattern and how dominant they are therein. The following classification of cues is useful.

Non-facilitative:	cues that neither inhibit nor intensify sexual arousal.
Facilitative:	cues that enhance but are not necessary for sexual arousal.
Paraphiliac:	cues that are necessary for sexual arousal

Thus we can distinguish inhibitory from non-facilitative, from facilitative, from paraphiliac cues. For example, a person who was turned off by physical or psychological trauma would not be a masochist. But an individual whose sexual arousal was neither inhibited nor facilitated by such trauma would be a non-facilitative masochist. An individual who could get aroused without such trauma but found being spanked, hard bites, or verbal abuse increased his arousal would be a facilitative masochist. Only an individual who, say, could only become aroused by having a stiletto heel ground into his testicles would qualify as a paraphiliac masochist. Similarly, we can talk of non-facilitative, facilitative, and paraphiliac fetishism, transvestitism, zoophilia, pedophilia, etc.

While there frequently is fairly close accord between one's sexual fantasies and one's arousal cue-response pattern ([20], p. 250), there is not always total overlap. Indeed, the disparity between the two frequently is large enough that whether to act out one's fantasies can be a difficult and even risky question (*ibid*., pp. 250–253). Just as we can classify arousal cue-response patterns as homosexual, bisexual, heterosexual, fetishistic, masochistic, etc., so too can we so classify our erotic fantasy structures (which by adulthood tend to be based on fantasy elements as stable as our cue-response patterns). And, similarly, we can distinguish between non-facilitative, facilitative, and paraphiliac ingredients in our erotic fantasy structures. Thus, a person who only can achieve arousal and orgasm if having masochistic fantasies (regardless of sensory stimulation) would be a paraphiliac masochist in fantasy structure, but a person who could enhance arousal with such fantasies but did not have to engage in such fantasies to be aroused and achieve orgasm would be a facilitative masochist in fantasy structure. At the same time, either such person could find the fantasized stimuli a "turn-off" in real life and so would not be any sort of masochist with respect to arousal cue-responses.

SEXUAL DISORDERS RECONSIDERED

With our understanding of the complexities of sexual identity and arousal now enhanced, we return to DSM-III's treatment of the paraphilias. However, since we have adopted the term "paraphiliac" as a part of our classification scheme, to avoid confusion for the most part we no longer will follow DSM-III in talking about the paraphilias; instead, we will refer to the behaviors it calls paraphilias as *variant sexual behaviors*.

Let us begin by returning to the fact that DSM-III allows preferred activity or fantasy of such activity to qualify one for suffering from zoophilia or pedophilia, but restricts fetishism, exhibitionism, voyeurism, and masochism just to overt activity and sadism to overt or simulated activity. Thus, an individual who cannot be aroused except by fantasizing intercourse with a dog but is unwilling actually to do it suffers from the mental disorder of zoophilia; but an individual who cannot be aroused except by fantasizing being a sex slave in bondage who is whipped but is unwilling to actually engage in such behavior does not suffer a mental disorder. Initially, it seems the difference here is quite arbitrary.

Is there any plausible rationale for including fantasy in zoophilia and pedophilia but not for the others? The only clue I can find in DSM-III is that "the paraphilias . . . in varying degrees may interfere with the capacity for reciprocal affectionate sexual activity" (p. 261), and thus in some cases mere paraphiliac variant fantasy behavior would interfere with the capacity for reciprocal affectionate sexual activity and in others it would not. As stated, the suggestion is not terribly plausible. In discussing animal contacts, Kinsey [17] writes:

> In some cases the boy may develop an affectional relation with the particular animal with whom he has his contacts, and there are males who are quite upset emotionally when situations force them to sever connections with the particular animal. . . . The elements that are involved in sexual contacts between the human and animals of other species are at no point basically different from those that are involved in erotic responses to human situations.
>
> On the other side of the record, it is to be noted that male dogs who have been masturbated may become considerably attached to the persons who provide the stimulation; and there are records of male dogs who completely forsake the females of their own species in preference for the sexual contacts that may be had with a human partner (pp. 676–677).

Thus neither paraphiliac zoophiliac fantasies nor activities need interfere with the capacity for reciprocal affectionate sexual activities – with animals. Similarly, the sexual relationships between pre-pubescent children and adults in some circumstances can be reciprocal affectionate sexual episodes, and so pedophilia – even if paraphiliac in fantasy or activity – need not interfere with that capacity.

These considerations suggest that for DSM-III it is the capacity for reciprocal affectionate activities with adult humans that marks the difference. Surely, if one's fantasy structure is paraphiliac, zoophiliac, or pedophiliac, it is going to be difficult for an individual to respond sexually with adult humans. For it will be difficult to have intercourse with an adult human while fantasizing it is either a sheep or a little girl; but it is not impossible – if one minimizes body contact with the woman to just genital contact (or straps fleece to her other points of bodily contact), it may be possible to fantasize she is a sheep and respond; and if one picks a small adult female with flat chest and shaved pubic hair, dresses her up in little girl dresses and "Mary Janes," it may be possible to copulate with her while fantasizing a pedophiliac episode. And in such cases, there would be capacity for reciprocal affectionate activities with adult humans despite the paraphiliac fantasy structures.[10] In reality, however, acting out such fantasies is unlikely to be satisfying in the long term and one expects it is unlikely that such persons will have high motivation for seeking out adult human partners. Thus, in these cases, there does appear to be a rationale for including paraphiliac fantasy structures in the diagnosis of zoophilia and pedophilia.

Does this rationale also provide a basis for excluding paraphiliac fantasy structures from the other variant sexual behavior classifications? It is possible to imagine how someone with paraphiliac fantasies of exhibitionism or voyeurism could arrange circumstances where with a willing partner one could incorporate reciprocal affectionate activities with paraphiliac fantasies. In the long term this may be difficult, since the fantasies are to be analogous to DSM-III's expectations that the strangers be unaware (voyeurism) or that there be no attempt at further activity with the "stranger" (exhibitionism). Yet masochistic fantasies of being "forced" to engage in unwanted sexual activity frequently are acted out in erotic psychodramas. So there is potential. Nevertheless, paraphiliac fantasy voyeurism and exhibitionism seem to have less potential for contributing to reciprocal affectionate activities with adult humans than does paraphiliac fantasy zoophilia or pedophilia. And with regard to fetishism, if one is a fantasy paraphiliac with respect to women's shoes the potential seems little better. Thus it appears that for most of the variant sexual behaviors, the potential for paraphiliac fantasies interfering with the capacity for reciprocal affectionate activities with adult humans seems on a par, and thus we conclude that the decision to include fantasy in some and exclude it in other paraphiliac clinical diagnoses in DSM-III is arbitrary under the reciprocal affectionate activities criterion.

The other side of the issue is to consider whether paraphiliac arousal cue-response patterns fare any better on the reciprocal affectionate activities criterion. To the extent that exhibitionism or voyeurism requires that the victims be unsuspecting strangers, the opportunities for incorporating such paraphilias into reciprocal affectionate activities are even less than for the paraphiliac fantasy forms where the person does not desire to actually engage in the behaviors. As noted above, with willing participants the prospects are better in zoophilia or pedophilia – providing non-human or pre-pubescent partners are allowed. In the case of fetishism, it largely depends on the partner's comfort with and ability to engage in fetishistic activity.

The crucial cases thus become sexual sadism and sexual masochism. For here we have the most complex interactions of possibly paraphiliac fantasy or real life conditions and also substantial potential capacity for reciprocal affectionate activities. As John von Neumann jokingly (but with considerable truth) put it, "A sadist is a person who is kind to masochists." There are significantly large numbers of individuals whose sexual fantasy structures enjoy high components of masochism or sadism, with the former being in preponderance; but there is no reliable data as to what proportion, if any, are paraphiliac in fantasy structure. With respect to arousal cue-response there appear to be significant numbers of persons who are facilitative masochists and somewhat fewer who are facilitative sadists. Whether there are any paraphiliac fantasy or arousal cue-response masochists or sadists is unclear, as there are no reliable data on the subject. Nevertheless, the available S/M guides or books with inside knowledge on such subjects (e.g., [10], [16], [37]) strongly suggest (a) that the overwhelming majority of persons actually involved in sadomasochistic sexual activity are facilitative masochists; (b) that much of S/M sex is psychodramatic acting out (simulation) among consensual partners, which minimizes actual physical or psychic trauma; (c) that even facilitative sexual sadists are in the minority; (d) the shortage of preferential participants in the sexual sadist role are such that persons preferring the sexual masochist role frequently, and often reciprocally, find themselves playing the sadist role to keep the psychodramatic S/M arena going. The basic picture gleaned from such insider writings is that S/M sex typically is, and involves high capacities for, reciprocal affectionate sexual activities, albeit in psychodramatic scenarios where such concerns are placed in simulated suspense.

These observations, coupled with the paucity of empirical research on sexual sadism, sexual masochism, and on the other "paraphilias," raise serious challenges. First, how many persons actually meet the criteria for being paraphiliac with respect to fantasy structure or cue-response pattern for any of the DSM sexual variations? So far as I can tell from the literature and counselling contacts, they are extremely few. Rather, the norm is for there to be quite a number who are facilitative in either fantasy or cue-response pattern, with substantially more of the former. Second, the commercial success of S/M paraphernalia shops and interviews with their proprietors suggest that there are substantial numbers of "couples" who are capable of addressing their facilitative S/M fantasies and working them into a consensual, cooperative, and affectionate sexual repertoire. My counselling experience reinforces this impression. In significant numbers of cases there is reason to believe that couples consisting of one facilitative transvestite can do the same. However many couplings break up over such fantasy structures and the expressed desire to act them out, there are substantial numbers of couples who can address, accept, and become comfortable with the need of partners to act out these fantasies and incorporate them into productive, reciprocal, and affectionate relationships. We have no data on whether these actings on variant sexual desires are any more or less a factor in the dissolution of relationships that potentially could be reciprocally affectionate than are other sources of relationship dissolution. Indeed, I suspect it is the basic quality of the interpersonal/love relationship that is more crucial to the success and survival of the relationship than the psychosexual idiosyncrasies of the partners.

VARIANT SEXUAL BEHAVIORS AS PSYCHOLOGICAL DISORDERS

With an appropriately subtle appreciation of human sexual identities, a classification scheme appropriate thereto, and an appreciation of the psychosexual subtleties involved, we are now in a position to reconsider the circumstances in which variant sexual behaviors should be classified as mental disorders and DSM-III's attempt so to classify them.

We must first address a serious problem – the paucity and low quality of research on variant sexual behaviors. The situation is rather similar to research on homosexuality in the mid-sixties and before, where there was a wealth of clinical data on mentally-disturbed homosexuals, some empirical research based on prison or psychiatric samples, and very little research based on "normal" samples of homosexuals. The portrait of homosexuality and its psychological correlates was highly distorted, being based on highly biased samples. Subsequent research based on non-clinical, non-prison samples radically changed our understanding of homosexuality and was an important factor in the removal of homosexuality from DSM. Research on other forms of variant sexual behavior catalogued in DSM-III is at a level no better than that of homosexuality research in the mid-sixties. There is an abundance of psychiatric literature presenting case studies and theorizing about the phenomena. To a surprising degree, the theories show striking resemblance to those erroneous psychiatric theories that attempted to explain the inevitable psychopathology of homosexuality. For example, Charles Socarides' treatment of fetishism [25] is based on pre-Oedipal identifications, as is his account of homosexuality [26]. Virtually all the empirical studies are based on samples of arrested sex offenders, and only for exhibitionism, transvestitism, and pedophilia have substantial numbers of studies been done. For all the variations there is a paucity of psychological studies. In the area of sadomasochistic sex, I know of only three empirical studies based on "normal" populations (one unpublished, [28] in German, and the other questionable); interestingly enough, these suggest that the psychiatric views of sadomasochistic sex are as distorted as were earlier views about homosexuality. Thus, DSM-III's treatment of variant sexual behaviors is on as shaky empirical foundation as was DSM-II's inclusion of homosexuality as a mental disorder.

These considerations suggest, but do not establish, that variant sexual behaviors should not be included in DSM, but that perhaps more restrictive ego-dystonic or disorders-of-impulse-control versions legitimately might be included. We now explore this idea.

The first question to ask is whether there really are paraphiliac instances of these disorders (in fantasy or cue-response)? We have no data that directly answer this question. Although "paraphilia" typically excludes homosexuality, given our use of "paraphiliac" we can talk of facilitative versus paraphiliac homosexuality in fantasy and arousal. In a large-sample study having the most representative homosexual sample yet, Bell and Weinberg [4] found that 64% of the white homosexual males had engaged in heterosexual coitus, and of those who engaged in heterosexual activity 92% sometimes or always reached orgasm (Table 3.4). Thus, the majority of their sample was not paraphiliac with respect to cue-response arousal.[11] With respect to heterosexual masturbatory fantasies, 23% of the white male homosexuals had them. Thus, while the majority apparently were fantasy paraphiliac homosexuals, nearly a quarter were not. These data, coupled with the previously-mentioned parallels between research on homosexuality and on the other variant behaviors, and my own sexual counselling experience, strongly suggest to me that persons who are cue-response paraphiliac with respect to DSM-III's variant sexual behaviors are extremely rare, if they exist at all; but I would expect there to be substantially more persons who are fantasy paraphiliac. The available evidence indicates that of those who engage in sadomasochistic sex the overwhelming majority are facilitative, not paraphiliac, in cue-response, but a substantial number may be fantasy paraphiliac.

Assuming that arousal cue-response paraphiliacs are so exceedingly rare, DSM-III's practice of not requiring that the variant behaviors be paraphiliac, but rather only preferred, begins to make some sense. For psychotherapists do encounter and attempt to treat persons for, e.g., exhibitionism, fetishism, pedophilia, etc. – either because the person is arrested for the behavior or else finds the behavior or fantasies ego-dystonic; neither of these circumstances requires a paraphiliac response.

In such cases it seems as appropriate for there to be a DSM classification for them as there is for ego-dystonic homosexuality. The issue is whether DSM-III's current classifications are appropriate.

Recall that DSM-III's conceptualization of mental disorders is that they are significant behavioral or psychological syndromes typically associated with *distress* or *disability* (impairment in one or more important areas of functioning as opposed to a conflict between an individual and society – p. 6). When the individual finds any of the variant sexual behaviors (or homosexuality or heterosexuality) ego-dystonic, then there is a case for including such behavior in DSM-III; it is the ego-dystonia, not its sexual source, which warrants inclusion, and so these should be on a par with non-sexual ego-dystonia – perhaps in a generalized ego-dystonic condition category. Whether such ego-dystonias should be mental disorders as opposed to including them among "Conditions Not Attributable to a Mental Disorder That Are a Focus of Attention or Treatment" is unclear for reasons discussed in Section I above. Since many of these variant sexual behaviors (e.g., transvestitism, sadism and masochism, fetishism, and many of the "atypical paraphilias") are not ego-dystonic to large numbers of participants, under DSM-III's conceptualization the only grounds for including them as disorders would be if they inevitably led to disability. Indeed, it was the conclusion that non-ego-dystonic homosexuality did not lead to such disability (as opposed to being a conflict between the individual and society) that led to its exclusion from DSM-III.

Do the variant sexual behaviors listed in DSM-III inevitably lead to disability? As discussed previously, DSM-III seems to view disability as involving impairment in the capacity for reciprocal affectionate sexual activity with adult humans. Is such impairment a disability as opposed to a conflict between the individual and society? A homosexual arousal cue-response pattern impairs capacity for reciprocal affectionate sexual activity with adult members of the opposite sex, but DSM-III judges that not to be a disability. In effect, the judgment is that society normatively has decreed that sex is to be confined to reciprocal affectionate activity with the opposite sex, and the homosexual who refuses is in a conflict with society. But for many male homosexuals, the conflict is not just over sexual object choice; numerous male homosexuals choose to separate their affectionate behavior from their sexual activity, confining the latter to impersonal, non-affectionate, non-reciprocal activity in orgy rooms at gay baths or in "glory hole" establishments. DSM-III evidently does not view such patterns of behavior as displaying disability either. Rather, the judgment seems to be that such homosexuals who reject the idea that sexual activity should be reciprocal and affectionate are just in conflict with society. Thus it appears that DSM-III cannot consistently view impaired capacity for reciprocal affectionate *sexual* activity as a disability in the case of the other variant behaviors, while denying it is for homosexuality. If the latter merely constitutes a conflict between the individual and society, then *prima facie* so too are the former.

In another article [32] I have examined in detail each of the variant sexual behaviors listed in DSM-III and the available data pertaining to them. That examination revealed that (a) the syndromes are not always ego-dystonic and thus do not always lead to subjective distress; (b) the syndromes do not inevitably constitute a disability any more than homosexuality does; thus, since the latter is insufficient to qualify as a mental disorder under DSM-III's "concepts", neither do the other variant sexual behaviors; (c) insofar as instances of variant sexual behaviors constitute a disability, it is in virtue of being compulsive or antisocial in ways that interfere with meeting other responsibilities. These are covered under existing categories ("Disorders of Impulse Control," "Antisocial Personality Disorder" [301.70], "Adult Anti-Social Behavior" [V71.01], that are not peculiarly sexual; (d) The variant behaviors constitute a conflict between society and the individual in ways strictly analogous to the situation for homosexuality. Those findings exploit the limited available empirical research evidence to conclude that none of the variant behaviors, even if arousal cue-response paraphiliac (and even less so if fantasy paraphiliac), automatically warrants inclusion in DSM-III as sexual disorders. In effect, the same considerations that prompted the removal of homosexuality and the attendant rethinking of the criteria for a mental disorder in DSM-III, if consistently applied, call for the removal of the other variant sexual behaviors. To be sure, the available research data is uneven, sparse, and not that reliable. But the data that plausibly support the inclusion of such disorders are even sparser and weaker. Absent reliable data based on non-clinical, non-criminal samples in well designed and executed studies, it is irresponsible to include such variations or even their paraphiliac versions in DSM-III; for there is no established factual basis supporting the claim that they meet DSM-III's criteria for being a mental disorder. Moreover, there is one excellent prison study [11] of arrested sex offenders, which found that their MMPI profiles did not differ significantly from those of the normative population for the MMPI – strongly suggesting there is no intrinsic connection between engaging in variant sexual behavior and possession of a mental disorder. Indeed, here and elsewhere [32] I have tried to show that the best available evidence lays a plausible basis for the conclusion that the variant sexual behaviors are not appropriately included, and that when manifestations of such behaviors warrant inclusion it is due either to ego-dystonias associated with them that are on a par with non-sexual ego-dystonias, or else they should fall under other antisocial behavioral or personality disorders or disorders of impulse. Indeed, other than for Gender Identity Disorders and various Psychosexual Dysfunctions (which I have not discussed here), it is unclear whether there are any *peculiarly sexual* mental disorders that cohere with DSM-III's conception of a mental disorder.

CONCLUSIONS

Our investigation of the so-called "sexual paraphilias" or variant sexual behaviors has reinforced the suspicion that they are not mental disorders, but rather constitute conflicts between the individual and society. Absent solid empirical research showing that such behaviors, even in their paraphiliac forms, meet the DSM-III criteria of mental disorders, their inclusion in DSM-III is unwarranted, unscientific, and only serves to reinforce the suspicion that in the psychosexual arena, at least, psychiatry reduces to the codification of social mores masquerading as objective science.

Committee on the History and Philosophy of Science,
The University of Maryland,
College Park, Maryland, U.S.A.

NOTES

[1] This is a much expanded and revised version of an invited paper presented to the Society for Health and Human Values in 1980. I am especially grateful to Robert Spitzer, M.D., for helpful discussions and comments. I have benefited from participation in, and access to, the working papers of a Hastings Center project on the APA controversy. Partial support for writing this paper came from the University of Maryland General Research Board.

[2] Unless otherwise noted, all page references are to the third edition, DSM-III [2]. DSM-II refers to the second edition [1], and ICD-8 and ICD-9 refer, respectively, to the 8th [40] and 9th [41] editions of the World Health Organization's *International Classification of Diseases*.

[3] Although feminist writings on the subject argue that rape is an act of violence, not a sexual act, and in support it appears to be the case that there is a high incidence of impotence on the part of rapists [19], it would appear that, whether rape is a sexual disorder or not, it ought to appear somewhere in DSM-III, which it does not. Similar comments apply to at least coercive incest.

[4] E.g., in various major cities in this country there are urophilia clubs where members hold meetings to engage in various "water sports," including urinating on each other and drinking urine.

[5] In DSM-III this was replaced by the "Ego-Dystonic Homosexuality" classification, which is discussed below.

[6] For details on such theories, cf. Suppe, [35]. Laws of quasi-succession tell how an external or input state and an internal state at times determine subsequent internal states. Thus, e.g., they can describe how an individual's current state and environmental influences determine the next state of the individual.

[7] It involves a college-aged satyr who has an exceptional sex-drive, has a history of working in a homosexual "call-boy service," does not respond erotically to males and only plays the passive/non-ejaculatory roles as a "call boy." Currently he is being supported by an older male "lover" who keeps him but only demands he fellate him. He is erotically very turned on by females but does not respond to males. In consultation he has reported efforts to become homosexual – on grounds that as a Gay it would be far easier to satisfy his enormous sex drive with other males; he reports severe frustration in failing to do so, and over the fact that – try as he may – he cannot become a homosexual. He currently is struggling with the frustrating circumstances of having to acknowledge he is heterosexual despite his strong desires and efforts to be homosexual and his fervent belief that he would be more satisfied and happy were he homosexual. He feels trapped in heterosexuality. If this is not a case of ego-dystonic heterosexuality, I do not know what one would be.

[8] Others such as Bieber and Socarides prefer to "cure the psychopathology." The issues of cure, including what counts as a cure, are complex and controversial. See my [31] for discussion of what constitutes a cure and the moral issues surrounding the attempted "cure" of homosexuality.

[9] Given this bias, it is surprising that extreme promiscuity does not show up as a psychosexual disorder or a paraphilia. Distress over it does appear under "Psychosexual Disorder Not Elsewhere Classified" (302.89).

[10] As the "Michael" episode reported in my [31] displays, the possibilities suggested here are no more bizarre than what some homosexuals have indulged in to effect heterosexual coitus.

[11] Their sample certainly is not representative, but it is more diverse and reflective of diversity among homosexuals than any other study yet done. See my article [30], pp. 72–75, for a discussion of their study and the representativeness of their sample. Included in their sample were persons who were bisexual in behavior; nevertheless, 74% of the white males currently were exclusively homosexual in their behavior. For simplicity I am reporting just white male data here; the figures are higher for black homosexual males and for females, except for female heterosexual orgasm, which was quite low.

BIBLIOGRAPHY

1. American Psychiatric Association: 1968, *Diagnostic and Statistical Manual*, 2nd ed., American Psychiatric Association, Washington.
2. American Psychiatric Association: 1980, *Diagnostic and Statistical Manual*, 3rd ed., American Psychiatric Association, Washington.
3. Bayer, R.: 1981, *Homosexuality and American Psychiatry: The Politics of Diagnosis*, Basic Books, New York.
4. Bell, A. and Weinberg, M.: 1978, *Homosexualities: A Study of Diversity Among Men and Women*, Simon and Schuster, New York.
5. Bieber, I.: 1978, 'Letters to the Editor', *Archives of Sexual Behavior* **7**, 511.
6. Bieber, I.: 1980, 'On Arriving at the APA Decision', Working Paper, The Hastings Center Closure Project.
7. Bieber, I., *et al.*: 1962, *Homosexuality*, Basic Books, New York.
8. Boorse, C.: 1975, 'On the Distinction between Disease and Illness', *Philosophy and Public Affairs* **5**, 49–68.
9. Broverman, I., *et al.*: 1970, 'Sex Role Stereotypes and Clinical Judgments of Mental Health', *Journal of Consulting and Clinical Psychology* **34**, 1–7.
10. Califfia, P.: 1980, *Sapphistry: The Book of Lesbian Sexuality*, Naiad Press, Tallahassee.
11. Carroll, J. and Fuller G.: 1971, 'An MMPI Comparison of Three Groups of Criminals', *Journal of Clinical Psychology* **27**, 240–242.
12. Coleman, E.: 1982, 'Developmental Stages of the Coming-Out Process', in W. Paul, J. Weinrich, J. Gonsiorek, and M. Hotvedt (eds.), *Homosexuality: Social, Psychological, and Biological Issues*, Sage Publications, Beverly Hills, pp. 149–158.
13. Crisp, Quentin: 1978, *The Naked Civil Servant*, New American Library, New York.
14. Monteflores, C. de and Schultz S.,: 1973, 'Coming Out: Similarities and Differences for Lesbians and Gay Men', *Journal of Social Issues* **34**, 59–72.
15. Engelhardt, H. T., Jr.: 1977, 'Is there a Philosophy of Medicine?', in F. Suppe and P. Asquith (eds.), *PSA 1976*, vol. 2, Philosophy of Science Association, East Lansing, pp. 94–108.
16. Greene, G. and Greene, C.: 1974, *S-M: The Last Taboo*, Grove Press, New York.
17. Kinsey, A. C., *et al.*: 1948, *Sexual Behavior in the Human Male*, W. B. Saunders, Philadelphia.
18. Kinsey, A. C., *et al.*: 1953, *Sexual Behavior in the Human Female*, W. B. Saunders, Philadelphia.
19. Groth, A. and Burgess, A.: 1977, 'Sexual Destruction During Rape', *New England Journal of Medicine* **297**, 764–766.
20. Masters, W., *et al.*: 1982, *Human Sexuality*, Little, Brown, Boston.
21. McCarey, J. and McCarey, S.: 1982, *Human Sexuality*, 4th ed., Wadsworth, Belmont.
22. Money, J.: 1980, *Love and Love Sickness: The Science of Sex, Gender Difference, and Pair-Bonding,* Johns Hopkins University Press, Baltimore.
23. O'Carroll, T.: 1982, *Pedophilia: The Radical Case*, Alyson Publications, Boston.
24. Shively, M. and DeCecco, J.: 1977, 'Components of Sexual Identity', *Journal of Homosexuality* **3**, 41–48.
25. Socarides, C.: 1960, 'The Development of a Fetishistic Perversion', *Journal of the American Psychiatric Association* **8**, 281–311.

26. Socarides, C.: 1968, *The Overt Homosexual*, Grune and Stratton, New York.
27. Socarides, C.: 1970, 'Homosexuality and Medicine', *Journal of the American Medical Association* **212**, 1199–1202.
28. Spengler, A.: 1979, *Sadomasochisten und ihre Subkulturen*, Campus Verlag, Frankfurt/Main.
29. Spitzer, R.: 1980, 'Homosexuality and Mental Disorder: A Reformulation of the Issues', Working Paper, Hastings Center Closure Project.
30. Suppe, F.: 1981, 'The Bell and Weinberg Study: Future Priorities for Research on Homosexuality', *Journal of Homosexuality* **6**, 69–97. Reprinted in N. Koertge (ed.): 1981, *Nature and Causes of Homosexuality: A Philosophic and Scientific Inquiry*, Haworth Press, New York.
31. Suppe, F.: 1984, 'Curing Homosexuality', in R. Baker and F. Elliston (eds.), *Philosophy and Sex*, 2nd ed., Prometheus Books, Buffalo, pp. 391–420.
32. Suppe, F.: 1984, 'Classifying Sexual Disorders: The Diagnostic and Statistical Manual of the American Psychiatric Association', *Journal of Homosexuality* **9**, 9–28.
33. Suppe, F.: 1984, 'In Defense of a Multidimensional Approach to Sexual Identity', *Journal of Homosexuality* **10**, 7–14.
34. Suppe, F.: 1982, Review article on R. Bayer, *Homosexuality and American Psychiatry*, *Journal of Medicine and Philosophy* **7**, 375–381.
35. Suppe, F.: 1976, 'Theoretical Laws', in M. Prezlecki, *et al.* (eds.), *Formal Methods of the Methodology of Science*, Ossolineum, Wroclaw, pp. 247–267.
36. Taber: 1977, *Taber's Cyclopedic Medical Dictionary*, 13th ed., F. A. Davis, Philadelphia.
37. Townsend, L.: 1972, *The Leatherman's Handbook*, Le Salon, San Francisco.
38. Tripp, C. A.: 1975, *The Homosexual Matrix*, McGraw-Hill, New York.
39. Tsang, D. (ed.): 1981, *The Age Taboo: Gay Male Sexuality, Power, and Consent*, Alyson Publications, Boston.
40. World Health Organization: 1968, *International Classifications of Diseases*, 8th ed., U.S. Government Printing Office, Washington.
41. World Health Organization: 1977, *International Classification of Diseases*, 9th ed., World Health Organization, Geneva.

JOSHUA GOLDEN

CHANGING LIFE-STYLES AND MEDICAL PRACTICE

Life is always changing and medical practice is always adapting to the changes. Historically it has been less apparent to us because the changes occurred slowly. The pace of modern change seems faster. Technological change, particularly, has made the dissemination of information almost immediate across the world. The consequences for medical practice are that health care personnel have to confront unique situations with ethical implications without having developed an ethical position to guide their conduct. The discussion that follows is intended to review some of the more significant and dramatic changes, which have sexual implications in the practice of medicine. Identifying the issues and discussing some of their sexual implications can help clinicians to begin to formulate their ideas. The always challenging task of determining appropriate action in the face of conflicting ethical principles can never be resolved by formulas. Each situation is unique because each person deciding what to do about it is unique. The other contributions to this volume have touched on many of the same issues. The focus on life changes and their consequences for medical practice should make the inevitable repetition tolerable.

While there are various ways of categorizing the changes in life-style, one convenient way of doing so is to divide it into attitudinal changes and technological changes. There are many areas of overlay, and undoubtedly miscellaneous changes as well. Examples of attitudinal changes are the greater social visibility and social acceptance of homosexual life-styles. Examples of technological changes include the availability of surgically implanted penile prostheses for men unable, consistently, to get and maintain an erection. Examples falling into the miscellaneous category include the economic pressure resulting from current political realities forcing medical practice into becoming more cost effective and therefore, inevitably, more oriented toward maintenance of health by prevention of disease.

Earl E. Shelp (ed.), Sexuality and Medicine, Vol. II, 137–154.

ATTITUDINAL CHANGES

Perhaps the most apparent recent rapid change is the change in perceived public attitudes toward homosexuality [6]. The Gay Liberation movement spawned in the last two decades has made homosexuality a not necessarily pathological variant of the conventional heterosexual life-style. It has also brought political power and economic influence to bear in the struggle for social acceptance of gay men, lesbian women, and other sexually different minorities in our society. The consequences for medical practice are numerous and important.

The board of directors of the American Psychiatric Association decided in December, 1973, that homosexuality, per se, was not a disease. That has not stopped a flood of patients coming to doctors for cure of their homosexuality. They rarely came to doctors for that reason before. If homosexual patients came to physicians at all, it was for more conventional ailments. Ordinarily, patients would try to disguise their sexual preferences because of their expectation that physicians would be discriminatory, treating them disrespectfully because of their socially conditioned attitudes of fear and loathing that comprise homophobia. Since all of us, homosexual, heterosexual, and intermediate categories, suffer from exposure to homophobia, doctors are no less likely to feel it than their patients. Horror stories about bad treatment from physicians are understandably common in the accounts of homosexual people. Many homosexual patients avoided physicians, even when ill, to evade the anticipated pain of dislike or rejection if their sexual preference were revealed. Avoiding medical care when it may be needed is dangerous, regardless of what motivates the avoidance.

Gay and lesbian people, many of them health care professionals, now teach medical students that they should not approach their patients with a heterosexual bias. Instead of referring to one's "husband, wife, or spouse" when taking a history, one should learn to refer to one's "lover or partner, or significant other." Instead of assuming that a sexually active woman needs contraceptives, it is better to ask whether the woman's sexual activities expose her to a risk of pregnancy. Although sexually transmitted diseases have always been a medical concern, the recent prevalence of Acquired Immune Deficiency Syndrome with its potentially fatal consequences makes it more important than ever that physicians know about the sexual practices of their patients. If a physician is to be able to help a gay male patient avoid a fatal disease like

AIDS, the physician must first know the patient is gay. Then, learning about which sexual activities are part of the patient's repertoire, the physician can advise him about risks, prevention, and treatment. If the patient fears rejection or disapproval of his life-style from the physician, it may cost him his life. Avoiding needed medical care deprives the patient of access to potentially life-saving information.

Implicit in the physician's role as an effective health care provider for homosexual patients is the physician's need to prepare himself adequately to talk openly about sexual issues with patients. Physicians, like most members of our sex-negative culture, tend to grow up somewhat sexually inhibited. Speaking openly and clinically about details of our own or someone else's sexual life is not generally done. Medical sex education is inconsistent. Many, maybe most physicians do not learn or later practice how to talk about sexual concerns with patients, families, and significant others. If they are shy, embarrassed, anxious, and awkward talking about sex, physicians will tend to avoid doing so. The ethical conflict here is straightforward. The physician is likely to compromise the patient's health and well-being for the physician's comfort and prejudices. Experience has taught most medical practitioners that they can learn to become comfortable speaking to patients about many other intimate details of their lives that do not ordinarily come into polite conversation. It is important for physicians to know that practicing the skills of sexual history-taking can make them comfortable and competent with the task. Only then, knowing what is possible, can they decide the ethical issues fairly.

Previously, the point was made that homosexual patients rarely seek medical help to treat their sexual orientation or preference for same gender sexual partners. However, parents of homosexual children and spouses of homosexuals more frequently seek a physician's help in trying to change the homosexual into a heterosexual. It is important to identify who the patient is. Usually, the "patient" who is experiencing the pain, anxiety, and related distress is not the homosexual man or woman the doctor is supposed to change. It is the loving, guilty, shocked, and perhaps disgusted parent or spouse who has to confront and understand his feelings about his loved one's homosexuality. Despite the recent changes in society's attitudes toward homosexuality, it is still true that homosexuals are often treated with disdain, rejection, and fear, as well as discrimination in many areas of life. Being homosexual in a homophobic society is hard, and most people, given a choice, would opt to be in the socially more acceptable heterosexual majority. Surprisingly, therefore, very few homosexual men and women seek to change their sexual object preferences and their homosexual life-style and cultural affiliation.

There is conflicting opinion [16] that most homosexual men and women can be changed into heterosexuals. Physicians need to know that. Nonetheless, parents, spouses, and others close to and caring for homosexual people are likely to seek a doctor's help in changing the homosexual. Those distressed patients need the physician's understanding, knowledge and support. They can benefit from learning that most homosexuals have little or no choice about being homosexual. They want to be relieved of guilt and blame over their child's or spouse's sexual preference. They need to learn that their acceptance and understanding is therapeutic, but their efforts to change the homosexual are usually not.

SEXUAL VIOLENCE

Another change in our view of sexuality has been the changing attitude toward sexual abuse. The liberalization of political thought coincident with the Civil Rights movement of the sixties and the reaction against United States involvement in the Viet-Nam war supported women's rights and reactions against violence. That led directly to developing social and political forces against rape and other forms of sexual coercion. Rape, which was a hidden crime, rarely reported, has become, unfortunately, more prominent in public awareness. This is due to more rapes being committed, particularly in industrialized societies, and to more of those rapes that are committed being reported. While the causes of rape and other forms of sexual abuse remain obscure, the effects are well-studied [8]. The dramatic short-term consequences of violent, sexual assault are reasonably well known. The delayed long-term effects on victims are currently being elucidated; they tend to be less well appreciated by society in general and by medical practitioners in particular.

The immediate consequences of sexual violence include physical trauma, possible infection, possible pregnancy, and inevitable psychological damage. Victims of sexual assault blame themselves and are blamed by others for being victimized. Some physicians are trained to treat the patient for the physical trauma they experience. The long-term effects of sexual assault – the persistent fear, sense of diminished self worth, violation, and sexual aversion some victims develop are usually neglected. Those adverse reactions are preventable in part; but physicians do not often anticipate potential problems and they tend not to treat complaints unless patients bring them up. The consequences of sexual abuse, shameful and degrading experiences for most victims, are not usually the complaints patients bring to their doctors.

CHILD SEXUAL ABUSE

A related issue that has gained prominence in the past three decades has been the problem of child abuse [11]. The initial emphasis placed on the problem was to identify the patterns of physical injury of abused children. The injuries were dramatic and the focus was on physical signs. With time, the emphasis has progressed to concern with the psychological effects as well. It is generally agreed that many abusing adults were themselves abused as children. The pattern of responding as victims to abusing, controlling adults continues into adulthood for many mature victims of child abuse.

When Freud [5] began his work on understanding mental illness, using the developing techniques of psychoanalysis, he learned of many instances of incest, sexual molestation of children, and other forms of child sexual abuse. He subsequently came to doubt that it could have happened so commonly that advantaged, educated, respectable parents, usually male, could do such heinous things. He therefore theorized that his patient's reports of sexual victimization were fantasied, not realistic, historical events. It now seems, in retrospect, he was wrong. As society's attitudes have changed, sexual acts against children ranging from incest to the use of children in pornography have been recognized as being more commonplace.

Many states now have laws in their medical practice codes requiring physicians to report cases of suspected child abuse, whether physical, mental, or sexual. I do not know of any reliable studies that address the issue, but I feel that most physicians ignore the law. They do not have a high index of suspicion and they would be reluctant to report parents or other suspected abusers if they did believe abuse occurred. The reasons are complicated. Despite accumulating evidence that treatment for child abusing adults is not effective [18], physicians shrink from the alternatives. Those involve disrupting families by removing children from parents and placing them in foster homes or juvenile facilities. Perhaps the same fears of breaking up a family that effectively coerce child victims, their siblings, and usually mothers into accepting chronic sexual abuse of children also influence doctors to ignore the laws. They may not want to believe abuse so generally condemned could actually occur. They may not want to get involved. Legal matters are disruptive, time-consuming, and emotionally stressful for doctors as well as patients. Fear of lawsuits may be mitigated in the many states which have laws protecting those who report suspected abuse from prosecution. Those protections do not seem to be sufficiently effective. For whatever reasons, and there may be others, the legal and ethical obligation that society imposes on physicians to report abuse has been of limited effect.

SEXUAL LIBERATION

Certain life-style changes, reflecting attitudinal changes about sexual behavior, have come to be referred to, loosely, as "sexual liberation." Included within the meanings of the term are such disparate phenomena as more unmarried adults cohabitating, children engaging in masturbation and coitus at earlier ages, and a more accepting attitude toward sexual relationships occurring outside of marriage between partners married to or committed otherwise to someone else. Each of these phenomena will be briefly examined with consideration of their implications for medical practitioners.

Recent studies document some of these changes [1]. For females, the age of first premarital coitus is lower than it has been in preceding generations. Females and males are experiencing sexual relationships with more partners and at an earlier age than recent past generations did. Associated with these observations, possibly causally related to them, there is a larger pool of sexually experienced peers available. There have been technological advances in the availability of contraceptive devices, and despite a raging ethical and political controversy, in the availability of abortion. Yet, the rates of premarital pregnancy keep rising ([2], [7]).

The ethical implications for medical practice are fascinating and compelling. It should be apparent that more frequent sexual activity at an earlier age probably means that children are exposed to the risks of sexual activity with less education, judgment and ethical certainty than might be desirable. It seems unlikely that most parents encourage or condone early sexual activity for their children. The pressures come from the children's peer groups, and from other societal forces that may sanction early sexual experimentation. Those forces may be the media, they may be parental behavior, they may be other less regularly condemned but necessarily changing social institutions like the influence of religion, for example. The result is that we have sexually active children who are more ignorant and less responsible about their sexual activities than would be desirable. The problem is, who will educate the children about sex? Parents do not. Schools do not. Physicians do not, unless they are asked, and then most do not do it anyhow, or do it badly because they feel unprepared. We are not even clear about what the education is for. Opponents of sex education have argued that if children learn about sex, they will do it. They do it anyway. If the schools, for example, were to teach about values and sexual conduct, it might interfere with the teaching of parental values. But parents predominantly do not teach about sexual values and conduct, at least not directly or effectively. The lack of parental influence may create a vacuum, but other factors fill it. There are powerful influences, like the peer group, and societal values expressed in the media, which affect the child's sexual conduct.

The results of sexual ignorance are unfortunately numerous and easily identified. Children born to young, often unmarried mothers have a greater likelihood of birth damage, emotional, and psychological deprivation, child abuse, limited opportunities for education, and the despair that is associated with all of the above. There is a cost to the well-being of the mothers whose lives are altered by the limitation on their opportunities for education, development, and support. There is a cost to society in terms of providing services to take care of the human suffering caused by unplanned, usually unwanted pregnancies happening to unprepared, unwilling children.

At a less dramatic level there is a cost to all human beings whose suffering from sexual ignorance is expressed in terms of sexual maladjustment. People who are unable to participate in and enjoy sexual intimacies are more likely to have difficulty establishing and maintaining a relationship. Marriages break up because of sexual dysfunctions. Children of those marriages suffer. People are anxious, depressed, defensive, and often isolated because of their failure to learn how to be sexually competent and comfortable.

PREVENTION OF SEXUAL IGNORANCE

Medical practice has a role in trying to treat the consequences of sexual ignorance. It may have a more important role in trying to prevent it. Ethically, medical practitioners are confused about what to do. The traditions of medicine have been to prepare physicians to treat illness in individual patients in their consulting rooms or hospitals. Economic forces and social practices have changed the earlier tendencies for clinicians to see patients in their homes. The medical care reimbursement system rewards, and therefore encourages, the practice of treating disease, usually advanced, in the most expensive technologically complicated setting – the hospital.

Although the economic pressures of the past few years have accelerated the changes in health care delivery, and although payment for medical care may predominantly be on a capitation basis, hoping to control costs, the preventive orientation has not yet permeated into the ethical sensibilities of most physicians. They may not see that it is desirable for them to prevent illness, in this case sexually-related illness, by taking an activist position in educating patients and the larger society. It will still be a long while before preventive efforts are rewarded by direct reimbursement. Most physicians will not approve of altering the physician's role to one of social activist and educator, yet there are some encouraging precedents, such as the physician-inspired movement to limit nuclear weapons. That movement came from a realization that the adverse health consequences of nuclear warfare were known to be much more serious than the less sophisticated lay public realized. In like fashion, the consequences of sexual ignorance are more serious than our passive indifference would indicate. Most physicians realize that. If an ethical obligation arises from that realization, then medical practitioners will respond by learning to be sex educators and counselors.

CHANGING FAMILY STYLES

Some changes in life-styles tend to be brief, dramatic, but of questionable significance for medical practice. Any change can be alarming, particularly in the area of sexual practices, because of the exceptional valence sexuality has in our culture. Two related phenomena are the experiments in communal living of the past two decades and the increasing tendency for adults to cohabit in intimate relationships over time without marrying ([3], [4]). There have been experiments in social organization of living groups throughout history. The nuclear family, consisting of a mother, father, children, and possibly grandparents or other relatives, has been the predominant model, and therefore reified as the norm. Departures from that norm have included various forms of communal living, as in "utopian communities," and polygamous families with a father and several wives and children. The current forms of social experimentation with family structure are as varied as they have always been, but some patterns are more popular. Households in which the parents are unmarried, but remain together over time and raise children, is one of the most rapidly growing patterns. Another is the single-parent family, either as a result of choice or loss of a partner through disagreement, death, or divorce. Other arrangements of particular interest are the increasingly numerous households of homosexual couples who want children, either conceiving with the help of a non-parenting partner, through artificial insemination, or who obtain children via adoption.

The communal arrangements are of various types. Individuals and "families" live together in organizations ranging from "cults" and religiously or philosophically based groups to loosely-structured, domestic (often rural) units, living together as friends who shared common interests and a belief in non-traditional values and living styles [4]. They often have different sexual practices and beliefs that may shock and frighten those whose sexual mores are "mainstream." There may be sharing of sexual partners outside of a relationship defined by either marriage or commitment.

The major implications for medical practice are curiously of a negative sort. That is, while many "traditionalists," including many physicians among them, believe any departure from traditional norms of family structure and sexual behavior are dangerously destructive of social order and morality, there is little evidence to support that belief. Children are born, raised, and seem to function normally, as do the children of traditional two-parent nuclear families, despite the differences in child-rearing attitudes and practices. The unmarried couples often seem to be as happy or unhappy as married couples. They may experience less unhappiness than couples committed to marriage, because when the inevitable conflicts associated with intimate living arise, they may separate rather than struggle to resolve their difficulties. The evidence provided by researchers about the psychological effects of alternative living styles on children and adults suggests no major pathology. Medical practitioners must learn about these variant life patterns in order to reassure themselves that their patients representing "variations" do not need special services or consideration because of being merely different.

ATTITUDINAL EFFECTS OF POPULAR CULTURE

There is a rapid change in the availability of sexually explicit material through the media. Films, magazines, television, books, and videocassettes are the most obvious examples of commercially available, sexually-oriented information provided to a hungrily consuming public. The public is getting educated. While there are justifiable doubts about the accuracy, tone, or esthetic sensibility of the materials available, there is no doubt about their popularity. The demand for sexual information and the public response to societal requirements for sexual competence fuel the advertising industry. Sex sells everything from fashion to political candidates to motorcycles. The primary motivation for the use of sexual information and titillation is mercenary. Making people sexually healthier and happier is not a primary motivation for the producers of sexually oriented materials. As a result, much of the information aggravates longstanding problems in the area of human sexuality. Myths about sexual performance are reinforced by making people aware, mistakenly, of such goals as multiple orgasms for women or "extending orgasm" for men and women. The competitive, goal-oriented culture that predominates makes more and better sexual functioning a desirable goal for most of us susceptible to cultural influences. Unfortunately, a sexual orientation toward performance is not conducive to good sexual performance. A pleasure orientation is.

By creating unrealistic sexual expectations, the availability of sexually more explicit information creates many problems, while solving others resulting from ignorance and lack of sexual information. One beneficial effect is the broadening of the concept of sexual normality. While everyone has wanted desperately to be sexually normal, in the past people have been at a loss to know precisely what that might be. People have suffered from feelings of inadequacy, believing themselves to be abnormal, when realistically their attitudes and practices were well within the broad parameters of what is believed to be sexual normality.

Publication of surveys on the sexuality of large populations has offered reassurance and comfort to many. It has also perhaps offered permission and encouragement to try new sexual behaviors, with as yet undetermined consequences. Examples are more sexual infidelity, early age of sexual activity for children, and more exotic practices like anal intercourse.

The effects of these changes on medical practice are variable. As sexual behavior becomes more of a concern, and is more of a matter of public discussion, people will turn to physicians for help with problems as they always have ([12], [13]). There will be a demand for information and guidance about sexual activities. Presumably, physicians will respond to that demand and will need to be adequately prepared to do so effectively. A proliferation of physicians and a continuing shortage of money to pay for health care will mean that physicians will have to compete for patients and offer services that patients want. They will not be able to be successful in economic terms while ignoring areas of patients' demands of which they disapprove or where they feel uncomfortable.

SEXUAL HEALTH CARE NEEDS OF ILL AND DISABLED PATIENTS

While the development of sexual therapy has been a recent phenomenon, many non-medical therapists have provided sexual health care. This is particularly true for the physically and mentally healthy, able-bodied segment of the sexually troubled population. One important effect of the sexual revolution is the increasing realization of the importance of sexual functioning for the physically and mentally disabled segments of the sexually troubled population. That may mean that medical practitioners will face more demands for sexual health care services from their patients and from the sexual partners of those patients. For example, breast cancer, which afflicts one in eleven women in the United States, is an illness with adverse sexual effects, touching the lives of an enormous number of women, as well as their sexual partners, and many of their relatives, friends, and children who have no breast cancer. That is only one relatively common illness. Since all illnesses have mostly adverse sexual effects, attention to the preservation of sexual functioning is important to the care of almost all patients and their significant others. The perception of sick people as "not sexual" has changed, and the implication for medical practice is clear.

Almost all people who choose to be sexual can benefit from efforts to help them achieve some means of sexual relating despite being labeled as ill or disabled. Medical practitioners will hopefully respond to the need. It may mean that doctors will be encouraging and instructing patients in sexual practices that were considered perversions and punishable by law just a few years ago.

TECHNOLOGICAL CHANGES

Changes which follow technological advances also influence medical practice. Transsexualism and surgical treatment for the condition are carefully discussed elsewhere in this volume. Another change is the increased availability of penile prostheses. These are surgically implanted devices that make it more easily possible for men unable to get or maintain an erection to achieve vaginal penetration. The use of penile prostheses to treat erectile dysfunction, usually for physically-caused, but occasionally for psychogenic reasons, has been well publicized. Many doctors and more patients are attracted to the possibility of a simple, mechanical solution to the complicated problem of a man's inability to insert his penis into a vagina.

There are many issues complicating the simple appearing solution. Firstly, most physicians are male. They have a male concept of sexuality which they share with most male patients. They believe "sex" is putting an erect penis into a vagina. They assume that most women share their belief, but many do not. Current evidence ([13], [9]) suggests that most women capable of orgasm do not respond with orgasm to the stimulation of penile thrusting in a vagina. Many women believe "sex" is tenderness, closeness, and more general caressing, as well as achievement of orgasm by various means. Second, since it is not common for physicians evaluating prospective males for penile prostheses to evaluate their female partners also, the physicians may not be aware of other conflicts or reservations that make the women reluctant to accept the prosthetically stiffened penis.

There are resulting ethical conflicts. Do physicians recommend prosthetic surgery for patients who may not be able to use the prostheses? The implication is that medical practitioners may have to expand their concept of who is their patient, if they are to offer good quality medical care. They probably should consider not only the identified male with erectile problems as their patient, but his sexual partner as well. It means spending more time in evaluation, learning the skills of evaluation, and devoting time and energy to following the two patients post-operatively to influence the outcome positively.

A related technological change in sexual behavior is the use of various mechanical devices to enhance sexuality. These devices range from a stream of water directed usually at the female genitalia, through the use of vibrators applied to genitalia, to dildoes, lubricants, lotions, and perhaps, at the edge of social acceptance, hard-core pornography and the accoutrements of sadomasochism – whips, chains, leather, and other erotically stimulating devices. While sexual aids of various kinds have been sought and used throughout history, the greater public discussion of sexual practices has made current use of sexual aids more apparent. The desire of some individuals to enhance their sexual response, and that of their partners, may present problems for medical practitioners.

Physicians, viewed by their patients as being wise authorities, may be called on for advice or permission either to use or refuse sexual enhancers. A familiar scenario is for a couple to present themselves to a physician to have him resolve a conflict between them. One member of the pair may want to engage in a sexual practice or use a device the other objects to, usually because it violates his sense of propriety. The physician may be asked to side with one against the other, supporting what are essentially ethical positions with medically based facts. The physician may say, for example, that oral-genital sexual contact is not necessarily harmful to health. He may agree that orgasm as a way of relieving sexual tension is acceptable, and that a vibrator may make it easily possible to achieve orgasm. He might agree that the restrained and controlled enactment of sadomasochistic fantasies by consenting partners would not be physically dangerous. Since the reasons for one partner's reluctant participation are rarely medically based, the physician's failure to respond to the underlying and unstated ethical objections may not help the couple resolve their conflict.

The ethical issues are further complicated by the potential differences between the values and sexual mores of the physician and those of one or both of his patients. While absolute guidelines do not exist, nor would they apply to the unique problems of any given situation, it is generally appropriate for the physician to be wary of imposing his ethical standards on his patients ([15], [17]). If he feels unable or unwilling to treat his patients without their conforming to his ethical position, he should, at the very least, refer them to competent colleagues who do not require such conformity.

If the issue is a simple one, where the physician is acting as an advisor and consultant to resolve an ethical conflict between only the sexual partners, some guidelines exist. Some jurisdictions, California, for example, have made sexual activities between consenting adults legal. Without the legal sanction, others have felt that any or all sexual activities, which do no harm (clearly a definition open to dispute), occur between consenting partners (able to consent freely and informedly), and occur privately or in ways that do not offend others – are acceptable, ethically. Without agreement between partners, the issue may need to be resolved by negotiation between the partners. Unfortunately, appeals to medical authority, medical fact, or ethical absolutes are rarely of help.

The sexual counseling role is a change in medical practice that broadens the scope of many physicians. Patient demand may require a physician to review the sexual system as routinely as he reviews the cardiovascular or gastrointestinal system in providing care. The physician as counselor in the resolution of ethical conflicts that may appear disguised as medical problems is also a broadened, somewhat new role for medical practitioners ([15], [17]). There will either be a need for physicians to devote more time to these areas than has been customary in the past, or physicians will need to refer patients to colleagues more willing and better able to meet patients' needs. In either case, more training for some physicians and some training for all physicians in the area of ethical, sexual conflicts are desirable.

SURROGATES

Another modern phenomenon, reflecting both attitudinal and technological changes in life-styles, is the use of surrogates for roles in sexually-related activities. The most common examples are surrogate sexual partners who collaborate with sexually dysfunctional patients. More recently publicized have been surrogate mothers who are inseminated by a father whose partner is unable either to conceive or to carry a pregnancy to term. Some of the surrogate mothers carry the pregnancy to delivery of a full-term infant and give the child to the father and his wife. The latest technology allows for some surrogates to be inseminated, and after a conception has occurred, the fertilized ovum is washed out of the surrogate mother's uterus and implanted in the uterus of the father's wife who could not conceive or implant, but who can carry the pregnancy through to delivery.

Ethical concerns exist about both surrogate sexual partners and surrogate mothers. The rationale for using them has always been that no more appropriate or suitable wife or sexual partner is available or able for whatever reason to function in that role. The use of surrogates is presently quite limited. Medical practitioners rarely confront the need to use a surrogate, since the practice is presently used by highly specialized individuals or teams of practitioners. The major problem, apart from the ethical concern of whether it is proper, is to determine the competence and emotional and psychological stability of the person functioning as a surrogate. The role is a difficult and complicated one. The emotional needs of the surrogate are inevitably going to influence the experience of working in a surrogate role. There is a need to balance the benefits that come from the surrogate's participation against the potentially adverse effects on the surrogate, on the sexual partner the surrogate is replacing, and on the other partner (usually a husband). There is also a need to consider the large context, such as the effects on the physician, the patient's family, and the larger society. A related issue, but one in existence longer and therefore seemingly generating less concern is the use of sperm banks for artificial insemination. Recently, a particular sperm bank has enlisted donors from the ranks of Nobel prize winners and other presumably intellectually superior, accomplished men to provide "eugenically superior" offspring to women seeking artifical insemination. While fascinating ethical questions ignite a controversy about the practice, its small scale makes it of no immediate, practical consequence for medical practice.

THE AGING POPULATION

People are living longer and the proportion of the population over sixty-five is increasing. As the population ages and some marital partners die, there are increasing numbers of older people without easy access to available sexual partners. Many of these older people will live in institutional settings like retirement homes, nursing homes, or convalescent hospitals. Coincident with this development has been an attitudinal shift in our culture. Just as we have come to recognize that physically ill and disabled people are sexual, want to be sexual, and benefit from help in restoring or maintaining their sexual functioning, so we have realized that sex is possible for and important to older people. That means that physicians may find themselves acting to help older individuals find sexual partners, frequently younger than they are. Older women, wishing to be sexually active, will be seeking partners among the ranks of younger men because there are too few sexually active older, single men to go around. If sex with partners, even casual partners, is not possible, then sexual activity may be confined to masturbatory, self-stimulation practices. Health care providers and health care institutions, which tend to change reluctantly, will find great difficulty in accommodating to the need to provide facilities and privacy for the fostering of sexual activities of an aging population. Medical practitioners will have to think through their biases and the ethical implications of what their role should be in promoting sexual health as an integral part of health care for an aging, but increasingly sexually active population. In this area of human concern, the ethical conflict requires a balancing of the contending interests of the aging patient, the busy, medical practitioner, and the health care institution with its own personnel and agendas. That does not make it unique. All ethical conflicts are generally more complicated than they appear to be before we think about them.

SUMMARY

History is dynamic and at times repetitive. Technological changes and attitudinal changes cause effects on one another. It seems to be characteristic of human beings to view their environment and culture in a narrowly focused way. We tend to overvalue the past, at times, and to ignore it at other times, discarding the value of much that has been learned all too painfully. In considering how medical practitioners adapt to the attitudinal and technological changes occurring in our culture, it is helpful to realize that technology always brings change. Sometimes it brings progress. Attitudes about sexual values change also, probably in response to ecological pressures and economic forces as well as technology. The long historical view of sexual attitudes and practices suggests that society alternates between cycles of permissiveness and openness on the one hand, and repression and inhibition on the other.

There is no reason to believe that our present attitudes are either permanent or deserve to be maintained.

Medical practitioners must recognize that sexual health is an important and meaningful part of general health. Presently, we appear to be in a cycle of sexual openness and experimentation. The effects on our lives will probably be much less significant than the doomsayers or the *avant garde* may claim. There are interesting changes occurring. Medical practitioners will consider them and utilize their abilities to include sexual health as part of what they do to help patients experience life positively. Technology will bring more changes, at an ever accelerating pace, and we need to think about the implications for sexual health care. The historical association of sexuality as more of a moral issue than a medical one is altering dramatically. The present still requires careful consideration of the ethical issues facing medical practitioners as we move from one attitudinal universe to another. This chapter addresses those ethical issues while identifying some of the more obvious changes.

Neuropsychiatric Institute,
Center for the Health Sciences,
Los Angeles, California, U.S.A.

BIBLIOGRAPHY

1. Chilman, C. S. (ed.): 1979, *Adolescent Sexuality in a Changing Society*, DHEW Publication (NIH) 79–1426, Washington.
2. Chilman, C. S. (ed.): 1980, *Adolescent Pregnancy and Childrearing*, DHEW Publication (NIH) 81–2077, Washington.
3. Eiduson, B.: 1983, 'Traditional and Alternative Family Life Styles', in M. D. Levine, *et al.*, *Developmental Behavioral Pediatrics*, W. B. Saunders Company, Philadelphia, pp. 193–208.
4. Eiduson, B. and Zimmerman, I. L.: 1984, 'Nontraditional Families', in L. L'Abate (ed.), *Handbook of Family Psychology and Psychotherapy*, Dorsey, Homewood, pp. 191–207.
5. Freud, S.: 1955, 'The Aetiology of Hysteria', in J. Strachey (ed. and trans.), *Standard Edition of the Complete Psychological Works of Sigmund Freud*, vol. 3, Hogarth Press and the Institute of Psychoanalysis, London, pp. 191–221.
6. Gagnon, J. H.: 1983, 'The Sources of Sexual Change', in G. Albee, S. Gordon and H. Lertenberg (eds.), *Promoting Sexual Responsibility and Preventing Sexual Problems*, University Press of New England, Hanover, pp. 157–170.
7. Gagnon, J. H. and Greenblatt, C. S.: 1978, *Life Designs*, Scott, Foresman, Glenview.
8. Hilberman, E.: 1976, *The Rape Victim*, American Psychiatric Association, Garamond/Predemark Press, Baltimore.

9. Hite, S.: 1976, *The Hite Report*, MacMillan, New York.
10. Kaplan, H. S.: 1974, *The New Sex Therapy*, Brunner/Mazel Publishers, New York.
11. Kempe, C. H.: 1979, 'Recent Developments in the Field of Child Abuse', in Ray E. Helfer and C. Henry Kempe (eds.), *Child Abuse and Neglect: The Family & the Community*, Ballinger, Cambridge, pp. 15–21.
12. Lo Piccolo, J. and Lo Piccolo, L. (eds.): 1978, *Handbook of Sex Therapy*, Plenum Press, New York.
13. Masters, W. H. and Johnson, V. E.: 1966, *Human Sexual Response* Little, Brown, Boston.
14. Masters, W. H. and Johnson, V. E.: 1970, *Human Sexual Inadequacy*, Little, Brown, Boston.
15. Oppenheimer, C. and Catalan, J.: 1981, 'Counselling for Sexual Problems: Ethical Issues and Decision Making', *The Practitioner* **225**, 1623–33.
16. Schwartz, M. F. and Masters, W. H.: 1984, 'The Masters and Johnson Treatment Program for Dissatisfied Homosexual Men', *American Journal of Psychiatry* 141 (2), 173–181.
17. Silbert, T. J.: 1982, 'Adolescent Sexuality: An Ethical Issue for the Pediatrician', *Pediatric Annals* **11**, (Sept.), 728–732.
18. Williams, G. J. R.: 1983, 'Responsible Sexuality and the Primary Prevention of Child Abuse', in G. Albee *et al.* (eds.), *Promoting Sexual Responsibility and Preventing Sexual Problems*, University Press of New England, Hanover, pp. 251–272.

NELLIE P. GROSE AND EARL E. SHELP

HUMAN SEXUALITY: COUNSELLING AND TREATMENT IN A FAMILY MEDICINE PRACTICE

Practitioners of family medicine provide "comprehensive medical care with particular emphasis on the family unit, in which the physician's continuing responsibility for health care is limited neither by the patient's age nor sex nor a particular organ system or disease entity" (official definition by American Academy of Family Practice). As such, they may be the only type of primary care physician who tend to have an opportunity to encounter patients with a full range of sexual complaints. Yet, family practitioners are inadequately prepared to deal effectively with these patients, since the curriculum in human sexuality is usually a token introduction to sexual anatomy, physiology, and pathology.

Sexuality is an integral part of one's being and personhood, expressed in one way or another in all that one does. It is a key part of one's wholeness as male or female, central to an understanding of who one is and how one relates to others (cf. [22]). The *sine qua non* of family practice is "the caring of the patient." In order to fulfill this mandate with regard to sexuality, it is important for family physicians to attain a more comprehensive knowledge of human sexuality. Although it may be difficult to expand curriculum time for this important subject, we recommend that practicing physicians educate themselves in human sexuality as keenly as they do other subjects. This brief essay demonstrates the importance of the pursuit by a discussion of several actual case vignettes.

Our purpose is to provoke a more sustained dialogue about the relevance of concepts of human sexuality to the routine practice of family medicine. One goal is to increase physician awareness of the implications to medical practice of changing life styles. In addition, and in the process, we shall illustrate how a physician's moral values can influence "medical" opinion. Finally, the case reports will illustrate the heterogeneous and complex nature of sexual complaints, and the possible

Earl E. Shelp (ed.), Sexuality and Medicine, Vol. II, 155–169.

difficulties of shaping a medically and morally proper response. We cannot be comprehensive, nor do we claim scientific "objectivity." Rather, we provide a clinically-based essay designed to stimulate reflection, evaluation, education, and research regarding the relationship of sexual concepts and moral norms to the routine practice of family medicine.

PRACTICE PROFILE

One of the authors (NG) has practiced family medicine for thirteen and one-half years. The first four years of private practice were in a suburban area of a large urban, English-speaking Canadian city with a population of approximately two million. The practice was primarily composed of heterosexual families. Approximately seventy-five percent of the patients were married couples with children. The remaining twenty-five percent were never married or divorced, separated, or widowed. The patient population was dominantly white and middle-class. There were approximately fourteen hundred active patients in the practice with between one and two percent known to be gay, lesbian, or bisexual.

The next eight years of practice were in academic medicine. As Director of Education in the Department of Family Medicine at Baylor College of Medicine, the author had few private patients but regularly saw patients in her teaching role as consultant and supervisor. Given the nature of this appointment, there were only minimal direct involvements with patients presenting with sexual problems. For the past one and one-half years this author has returned to private practice in Houston, Texas. The patient population is drawn primarily from the inner-city neighborhood where the office is situated. Patients tend to be upper income professional singles and couples with one child. The majority of patients are under age forty, primarily Caucasian, but other races and ethnic groups are represented with more women than men. The practice is located within one mile of a neighborhood in which a large concentration of gay and lesbian people reside. As might be anticipated, the practice seems to include a higher than expected percentage of gay and lesbian patients. Among the non-heterosexual, non-married patients a higher percentage of the men are homosexual than the women.

Initial consultations in at least ninety-five percent of the cases involve a complaint or perceived need for health maintenance: for example, in the area of sexuality, a PAP test or screen for venereal disease. Approximately six percent of the patient population describe themselves as non-heterosexual. During routine examinations, including the initial consultation, each patient is asked her or his sexual preference and if there are any complaints, worries, or questions that she or he would like to discuss. These inquiries are prefaced by a statement that these inquiries are part of the general evaluation.

The exchange tends to be freely undertaken and can entail the use of frank and/or slang terminology. For example, male homosexual patients may refer to their use of inhalants during intimate experiences by their street names rather than chemical names, i.e., "poppers" rather than amyl or butyl nitrate. The patients do not seem threatened or made uncomfortable by the initial inquiry or subsequent discussions about sexuality. The types of "treatment" or "problems" presented by patients include treatment and prevention of venereal disease, contraceptive advice, gynecological difficulties, pregnancy and its termination, stress and anxiety about sexual thoughts and behaviors, problems in relationships with the patient's partner, and many other issues. Given the scope of the examination related to sexuality, the authors' observation is that sexual "problems," either directly or indirectly, appear as prevalent among this population as everyday "headaches." If the physician asks about these matters, answers usually are forthcoming. For example, a routine question during the examination of patients is, "Do you have any difficulty with sexual intercourse?" It is not uncommon for the patient to respond, "Well, yes, now that you mention it." Often the patient will describe the difficulty. The final comment may be something like, "But, I never thought of it before."

We cannot be certain if the patient indeed never thought of it before or never thought that it would be appropriate to discuss this particular complaint with the physician without first getting the physician's permission to discuss sexual matters. The seemingly free and non-threatening exchange that characteristically takes place in discussions of sexuality with patients may be due in part to an absence of any indication of the physician's religious commitments in the waiting room, treatment rooms, or office. Care is taken not to disclose religious commitments, personal tastes, or values in conversation that may appear prejudicial or judgmental. We only can speculate what impact such a secularized encounter has on the patient and her or his willingness to discuss sexual issues. Our intuition is that the patients are unable to form prejudices regarding the views that the physician might have about any specific sexual issue. Accordingly, an environment is established that is permissive and neutral with regard to sexual topics that can be religiously volatile.

PRACTICE PHILOSOPHY

A basic task of the medical profession, in our judgment, is to make available to people their training, education, experience, and skills for implementation by patients of their life-plans. This objective is accomplished primarily by helping patients to have accurate and sufficient data on which to make informed choices regarding their bodies and "health."[1] Patients are seen as persons, individual and complex. They are not artificially compressed into disease categories. Rather, they are evaluated in as comprehensive a manner as possible (i.e., they are seen as biological, psychological, and spiritual beings). One premise for this approach is that "health" embraces more than an absence of disease.

A fundamental objective for the patient-physician consultation is a resolution of the patient's perceived problem. Some "problems" may not be perceived by patients (e.g., hypertension). When made aware of the "problem," the patient is encouraged not only to recognize that it exists, she or he is helped to see it as a sufficient threat to warrant compliance with recommended treatment. The encounter with patients is founded on the belief that the permission of the patient is necessary in order to allow the physician to participate in her or his life. This premise and principle stands in opposition to "strong" paternalistic behavior, but not necessarily in opposition to speaking and acting with authority during patient-physician interchanges. The author's preference is for patients to assume personal responsibility for their lives, and to be actively involved in therapeutic and preventive practices. An implication of this perspective is that medicine, in its most comprehensive form, is relational and technical. Technology is considered vitally important to competent medical care, but it is not considered an adequate substitute for the relational or person-affirming character of medicine. Technology is seen as a servant to human goals, including medical practice. The technical aspects of medicine, therefore, are seen in a servant role rather than that of master. The clinical encounter, including the multidisciplinary doctrines that are brought to bear within it, is an evolutionary and dynamic one that necessarily requires the physician to exercise discretionary judgments concerning which path to follow.

CASE DISCUSSIONS

This approach to the practice of medicine has several practical applications to patients with sexual complaints. Following a brief discussion of these general applications, six specific cases will be commented on to illustrate how this approach is expressed in actual situations. As mentioned above, a basic principle of practice is that patients are encountered in a relational and technical mode. With regard to human sexuality, this principle requires a response to patients that is sensitive to the emotional and functional elements of their sexuality. Too often complaints of patients are considered only insofar as they represent some form of physical dysfunction or some form of performance failure. Attention to this aspect of a complaint is important. However, the emotional conflict, distress, uneasiness, or anxiety that may attend a physical dysfunction or performance failure should not be overlooked. A perception of patients as integrated beings warrants this form of comprehensive approach. Since many patients have religious commitments that bear directly on human sexuality, the role that these views may have to free or burden a patient with regard to a sexual matter also warrants investigation. This is not to claim for the physician a spiritual authority. Rather, it is an attempt to be alert to every possible source of complaint and resource for its resolution. Appropriate referrals can be made if one decides that a patient may be helped by a religious authority.

Basic to the philosophy of the author's practice is a concern to involve patients in their own health care. Patients are encouraged to identify and define their problems as they understand them in consultation with the physician. These problems may be physical or emotional and very much influenced by the patients' life situation. For example, textbook definitions of sexual dysfunction would include impotency. However, impotency for a priest who has made a vow of celibacy might not be necessarily a problem that warrants medical intervention. One would want to rule out some organic basis for impotency; for example, diabetes that might have implications for the patient's total care. By allowing patients to participate in the identification of problems and by seeing patients as integrated beings within a social context, the role of the physician tends to be that of a resource to the patient in her/his execution of a life plan rather than an authority that imposes a pre-set view of the world upon them. The medical response becomes one directed toward the relief of expressed and/or discovered problems, to the degree possible.

The concern to see patients as integrated beings in a social context rather than in isolation assists the physician to respond in ways that can be most consistent with the patient's life style and should, in general, result in a higher level of compliance in the therapeutic regimen than might be obtained otherwise (cf. [21]). Patients are consulted from a position of value-neutrality. No judgments are made regarding their sexual behaviors as either right or wrong (cf. [8]). As individuals, each is affirmed as a being of value and worth. The neutrality of the physician, however, does not extend to the assessment of sexual behaviors, broadly understood, that are considered potentially physically or emotionally threatening. Opinions concerning these potentially harmful practices are communicated candidly to the patient with recommendations as to how to lessen the potential threat or to avoid it altogether. Patients are left free to accept the recommendation or not. Regardless of their choice, the attitude of neutrality regarding right or wrong is maintained. And, more importantly, each patient remains valued as a person. For example, homosexual male patients are counseled concerning the risks of acquired immuned deficiency syndrome (AIDS), along with suggestions as to how to lessen their risk to the extent this is known. If these patients choose not to follow the recommendations, they should surely understand that they are electing to run greater risks relative to their sexual behavior and necessarily assume the responsibility for it. This is not different from the personal assumption that heterosexual patients assume with respect to herpes, pregnancy, or other sex related "problems". The freedom of patients to order their risks is respected. The patient and physician recognize that certain disvalued outcomes (disease, unwanted pregnancy) may ensue, but these are considered a necessary cost of freedom.

The refusal to make judgments of right or wrong, while holding fast to valuing each patient, contributes to the maintenance of a constructive atmosphere in which mutual respect and trust are essential components. In the long run, it is believed that the rapport with patients is better, that patients tend to assume greater responsibility for their lives, and that the physician feels less pressure to categorize or "fit" patients into preconceived and possibly stigmatizing categories. Several case reports will illustrate the utility of this approach.

Case 1: Homosexuality, Artificial Insemination

Ruth is a 32-year-old lesbian who has been a patient for one year. She requested a screen for health on a male homosexual friend. He has previously fathered a child and was planning to be a sperm donor for her and her lesbian lover by self-administered insemination.

Ruth was seen in order to discuss this request in greater detail. Both she and her female partner are known to the physician. Ruth is stable, healthy, and has a good job. She and her lover Jane have been together for three years and seem to have a good relationship, flexible yet stable. Jane has a 6-year-old child from a previous heterosexual relationship who is doing well in kindergarten. Ruth and Jane appear fit physically, emotionally, and financially to be parents. Since it is unusual to undertake "tests for parenting" with heterosexual partners who request artificial insemination, it would seem a double standard to utilize such tests with a homosexual couple. Apart from the concern for fairness to the couple, other issues warrant consideration.

Perhaps there should be some special reluctance to comply with the request out of concern for the interests of the prospective child. However, studies of children in homosexual families have not been shown to be necessarily disadvantaged as a result of their family life in and of itself ([10], [11], [12], [18], [25]). Another concern is legal. Could the child bring suit against the physician if born with some form of anomaly? Can a physician guarantee the health of donor and child? Could the child sue for wrongful life for lack of a "normal parent" (cf. [25])? Could the donor at some point in the future claim fatherhood? What recourse would the couple have? What legal obligation does the physician have? Why not A.I.D. so that the donor is truly anonymous? Finally, a medical quandary. What would an adequate screen for health involve? A history, physical exam, and sperm study can screen to a certain extent, but more extensive laboratory evaluations would be better. What is the limit?

Several years ago this request would have been viewed as contrary to public morality. New information on these families and the effect on children within them seem to contribute to a greater medical acceptability of the practice. If A.I.D. is available to heterosexual couples, then homosexual couples should not be considered differently.

Case 2: Widowhood, Masturbation, and the Church

Mary, a very recent young widow and a religious fundamentalist, complained of anxiety and insomnia. Since her husband's sudden death in a motor vehicle accident, she has been unable to "function." She missed him terribly. She missed the sexual relationship. She finally volunteered during an interview that she had masturbated and is considering non-marital intercourse. She felt very guilty. She felt that she had let her church down. Her anxieties about her situation, sexual practices, and intentions were affecting her normal functioning. How should she be counselled?

The choice of living with anxiety and insomnia or living with guilt was a poor choice. It was explained to her that sexuality is an expression of health and of fundamental importance to human wholeness. Her current "singleness" and sexuality had to be reconciled within her present context. Had she considered extramarital intercourse while her husband was alive and had they promised fidelity to each other, the morality of the act would be questionable as a breach of promise. However, in her current single state masturbation and/or intercourse could be means of expressing and fulfilling her sexual needs and, hence, her health. She was told that masturbation is a common part of sexual behavior for individuals of all ages, infancy to old age. It is harmless to self and others, a way to enjoy one's body and release tension. However, many adults, raised with a negative attitude toward masturbation, still feel guilty about this form of sexual expression. Some prefer self-denial. Mary said that she had not discussed the conflict with the minister at her church. She said, "I would rather suffer than bring it up. We don't talk to him about sex." She was told that she did not need permission from the physician to perform the intended sexual acts, that her conflict was common, and that she had to choose between living with anxiety and insomnia or satisfying her sexual needs.

Case 3: Sexual Activity and Risks of Disease

A 27-year-old homosexual male with a history of previous venereal diseases, which include gonorrhea, syphilis, and venereal warts, requested counseling about how not to infect or be infected by his sexual partners.

In the past, no sexual relationship or a monogamous relationship would have been recommended. Such advice for this patient today is probably inappropriate and inconsiderate (cf. [1], [15], [16]). Having known the patient rather well, a suggestion of refraining from intercourse would have caused increasing depression associated with limited social life. This depression often affected his work performance and resulted in office visits with somatic symptoms. Which problem, depression or the risk of being infected again, is worse? He had previously requested his partner to use a condom. The patient's request was met with anger and rejection. It seemed that he had done all that he could in terms of prevention. It was his choice as to which risk to take. His sensitivity to the social consequences of venereal disease was admirable. And if he suspected an infection, he probably would be at the office right away.

Case 4: Middle Age, Sexuality, and Handicap

Martha, a 50-year-old homemaker, was seen in the office complaining of unfulfilled sexual desires. Her husband had been diagnosed with Alzheimer's disease two years ago. The family had been patients for five years. How could she be helped?

Sexual expression and fulfillment is a normal need at any stage of human life. Martha was assured that it was normal to have sexual feelings. She was questioned about the sexual relationship with her husband. She reported no contact. She really did not find him attractive anymore. She did not object to masturbation. However, she was not inclined to have extramarital intercourse, even though it had crossed her mind. She was asked about their commitment to each other concerning their sexual activity in marriage. If they had pledged monogamy, then extramarital sex would mean infidelity. If they did not pledge sexual fidelity, extramarital sex would have a different meaning. She reported that she had not discussed the conflict with her pastor for fear of being thrown out of the church.

She viewed sexual intercourse as an expression of love. She admitted that they satisfied each other before her husband's dementia. However, since he was incapable of satisfying her now, she felt that the satisfaction of her sexual needs should come from somewhere else. Martha said she felt that she was abandoning her husband if she had extramarital intercourse. She could not consider a sexual act as a fulfillment of a biological need similar to drinking water to quench a thirst. She was reassured that masturbation was medically normal. She was encouraged to continue working on her conflict.

Case 5: Changing Sexual Practices

A woman married five years sought help for severe insomnia and headaches. On careful questioning, she confessed that she had been worried about her husband's interest in group sex. She considered this activity "abnormal."

Her confusion and conflict were understood. She was reassured that her headaches and insomnia were most likely tension- and anxiety-related. She was asked about her marital relationship. She reported that her husband had desired some novelty in their sex life and thought that group sex was acceptable. He claimed to love her very much. He wanted her to expand *her* experience so that they could enrich *their* sexual experience with each other. She did not like that idea at all, and felt group sex was abnormal. She was told that group sex was not the usual and customary practice in our society. At the same time the meaning of "normal" was discussed. How is normal defined? Who makes the decision? What moral warrant does it carry? For a physiologic factor such as blood pressure, large population studies are used to set statistical parameters. But for sexual expression, establishing a non-evaluative meaning for normal is not easy. She was told that many years ago masturbation was considered abnormal and a disease by certain sectors ([16], [9], [23]) and that homosexuality, considered abnormal by some, was recently deleted from DSM III [1]. Views change over time as more knowledge of life experiences is gained. From a medical perspective, it seemed that their marital relationship was strained, that the sexual issue was just the ticket to the physician's office, and that treating her specific complaint without considering the total life-context would be doing the couple little good.

Case 6: Sexual Preference and Career Counseling

A young, ambitious, successful, and career-oriented heterosexual male model sought counselling to "see if I am crazy." He firmly believed that his career advancement was related to his ability to perform homosexually.

This was an unusual request for a first visit. The initial "gut reaction" was to refer him to one of the homosexual patients in the practice. However, it seemed important, on further reflection, to determine if the patient was covertly seeking approval to be gay. This was overtly not the case here, and if it were his life-style would not have been subject to condemnation. His career goals and life-plans were discussed. He perceived himself "out of place" in a career dominated by homosexual men. He was convinced that, if he could perform homosexually, his career would advance twice as fast. The dilemma was whether to refer him for career counseling or to explore further what fears, fantasies, and blocks he had. The latter course was followed. The patient felt that performing homosexual acts was indeed different from being a homosexual. He would do it if he could think of it as an act like breathing or swallowing. However, he was afraid that after repeated attempts he might like it, get involved, and become gay. No resolution was reached on this initial visit. The impression was that this patient may covertly be seeking approval to be gay.

DISCUSSION

These selected cases illustrate the sort of dilemmas that family physicians can face in counselling patients about sexual matters. In a recent study of the approach of family physicians to moral problems common to their practice, Christie *et al.* concluded that family physicians are reluctant to interfere with or influence patients regarding their complaints related to sexual life-style. Reasons cited for the physicians' discomfort included an absence of an effective technique for responding to these complaints and a fear of adverse reaction on the part of the patient ([4], p. 1137). In our judgment, this felt inadequacy and hesitancy on the part of physicians can and should be corrected in order to adequately care for those individuals who entrust themselves to family practitioners.

Large numbers of patients, as indicated by the above cases, seek clarification and/or approval from physicians for sexual behavior or desires. Patients worry, "am I 'normal'"? Family physicians, like marriage counselors, routinely deal with problems and confusions related to the many meanings given to the term "normal" and the difficulty of defining "perversion" ([13], [19], [24]), particularly for activities or fantasies generally considered socially unacceptable. Both terms, "normal" and "perversion," connote value judgments and corresponding attitudes that when used by physicians may appear as scientific judgments or medical truths grounded in objective research when, in fact, the term may only reflect the personal attitudes and value judgments of the speaker without scientific or medical warrant (cf. [20]). Few people dispute the claim that values pervade all of medical practice, including its so-called scientific findings. Thus, it is impractical, if not impossible, to urge physicians to be free of values. This sort of "mechanical" care-provider would be at variance from the moral tradition of medicine, and would probably not serve the best interests of patients. We are not calling for a devaluing of medicine. Rather, we merely want to underscore that, when medicine purports to describe reality, it is in fact reporting values and judgments that are more widely shared than competing values and judgments. Because of the general nature of these pronouncements, they may appear as brute facts rather than the values that they are [5]. In matters of human sexuality, the somewhat arbitrary and value-laden character of definitions of medical normality, abnormality, perversion, sick, and healthy are displayed prominently.

Society often asks physicians, individually or collectively, to perform a gatekeeping service and to function as a social conscience. This role of medicine is illustrated by the agency of physicians in the control of human reproduction, specifically abortion and artificial insemination. However, technical expertise to diagnose, prescribe, and treat complaints of patients does not necessarily entail conceptual certainty or knowledge regarding questions of "normalcy" or judgments of right and wrong with respect to a patient's sexual behavior or thought. Physicians as moral agents face quandaries in the formulation of their own codes in much the same way as non-physicians and society do. There is nothing inherent in the profession of medicine or the role of a physician that endows an individual with a superior moral understanding of right and wrong (cf. [14], p. 439).

Given the uncertain or, at best, relative nature of a physician's judgment regarding sexual morality, and given the rapid evolution of societal concerns about sexuality, how can family physicians competently and comfortably counsel patients about sexual questions or complaints? What attitudes should physicians embody? A complete response to these questions entails a discussion far beyond the suggestive character of this essay. However, as a beginning, we would caution against a too easy imposition on patients by physicians of personally held values regarding sexuality presented as medical "truths." To this end we propose that physicians refrain from labeling sexual activities as normal and abnormal, immoral, or suggesting that some sexual life–styles are better or more successful than others (e.g., more satisfactory, more enriching, more pleasurable). This is not to suggest that a family physician does not have an obligation to discuss with patients the potential physical and/or emotional risks associated with certain sexual activities. We suggest that physicians think of sexual life–styles as choices and, to the extent that they are choices, as aesthetic judgments. These life–styles are fashioned and implemented in terms portending a more rich, more beautiful, or more supportive structure for the values particular humans esteem. The life–styles are not ethical choices, in and of themselves, in the sense that they do not coerce others to participate or tolerate beyond the limits acceptable to a secular, pluralist society ([7], p. 116).

Perhaps physicians should see themselves as "value brokers." In this role they can (1) assist patients to perceive a range of values related to human sexuality, and (2) help sort out conflicts of value experienced by a patient as sexual choices are made. Third, we would urge that the physician *qua* physician during these encounters maintain a position of professional neutrality, to the extent necessary, respecting the individual rights of patients to select among moral values and moral options. Therapy is enhanced, in our judgment, when patients define or participate in defining a problem [17]. Similarly, it is appropriate for them to define a solution that is consistent with their values and aesthetic choices [2].

CONCLUSION

It is conceivable that the cases described above and suggestions regarding a physician's approach to sexual problems or complaints do not conform to the traditional therapeutic approach within medicine. So be it! Engelhardt is correct in his caution that "therapy" ought not to be restricted in meaning or form. He observes, "One can give therapy (treatment in the sense of care) to decrease difficulties associated with a sexual life–style, without aiming at eliminating (in a sense of curing) that life–style" ([7], p. 112). This approach is analogous to treating pain associated with menstruation, though not the menstruation itself. We suggest that the attitudes and skills articulated and implied above can be learned and should be taught during residency training for family physicians and other medical specialties likely to encounter patients with sexual concerns. The current literature in family medicine (e.g., [26]) alludes to the importance of an understanding of ethics and human sexuality. But little is said with regard to integrating or applying what is known about separate disciplines at the point at which their common interests are manifested (i.e., in the clinical encounter). In our judgment this artificial segregation and denial of the interrelationship and influence of one on the other can be sustained no longer. By recognizing the role of personal and social sexual values and their justifiable force in clinical practice, physicians should be more able to respond more appropriately as professionals and persons to complaints and questions regarding human sexuality.

Private Practice,
Houston, Texas, U.S.A.

Institute of Religion and
Baylor College of Medicine,
Houston, Texas, U.S.A.

NOTE

[1] See [3] for an extended discussion of concepts of health.

BIBLIOGRAPHY

1. Bayer, R.: 1981, *Homosexuality and American Psychiatry*, Basic Books, New York.
2. Bibace, R., *et al.*: 1978, 'Ethical and Legal Issues in Family Practice', *The Journal of Family Practice* **1** (5), 1029–1035.
3. Caplan, A. L., *et al.* (eds.): 1981, *Concepts of Health and Disease*, Addison-Wesley, Reading.
4. Christie, R. J., *et al.* (eds.): 1983, 'How Family Physicians Approach Ethical Problems', *The Journal of Family Medicine* **16** (6), 1133–1138.
5. Engelhardt, H. T., Jr.: 1975, 'The Concepts of Health and Disease', in H. T. Engelhardt, Jr., and S. F. Spicker (eds.), *Evaluation and Explanation in the Biomedical Sciences*, D. Reidel, Dordrecht, pp. 125–141.
6. Engelhardt, H. T., Jr.: 1974, 'The Disease of Masturbation: Values and the Concept of Disease', *Bulletin of the History of Medicine* **48** (Summer), 234–248.

7. Engelhardt, H. T., Jr.: 1980, 'Value Imperialism and Exploitation in Ethical Issues', in W. H. Masters, *et al.* (eds.), *Sex Therapy and Research*, vol. II, Little, Brown, Boston, pp. 107–137.
8. Farley, M. A.: 1978, 'Sexual Ethics', *Encyclopedia of Bioethics* IV, Macmillan-Free Press, New York, pp. 1575–1589.
9. Fortunata, J.: 1980, 'Masturbation and Women's Sexuality', in A. Soble (ed.), *Philosophy of Sex*, Littlefield, Adams, Totowa, pp. 389–408.
10. Green, R.: 1978, 'The Burden of Proof is on Those Who Say No', in J. P. Brady and K. Brodie (eds.), *Controversy in Psychiatry*, W. B. Saunders, Philadelphia, pp. 813–828.
11. Green, R.: 1978, 'Sexual Identity of Thirty-seven Children Raised by Homosexual and Transsexual Parents', *American Journal of Psychiatry* **135** (June), 692–697.
12. Hancombe, G.: 1983, 'The Right to Lesbian Parenthood', *Journal of Medical Ethics* **9**, 133–135.
13. Levy, D.: 1980, 'Perversion and the Unnatural as Moral Categories', in A. Soble (ed.), *Philosophy of Sex*, Littlefield, Adams, Totowa, pp. 169–189.
14. Linton, E. G.: 1972, 'Should the Physician be the Conscience of Society', in M. L. Silverman (ed.), *Marital Therapy*, Charles C. Thomas, Springfield, pp. 435–441.
15. Marmor, J. (ed.): 1980, *Homosexual Behavior*, Basic Books, New York.
16. Masters, W. H. and Johnson, V. E.: 1979, *Homosexuality in Perspective*, Little, Brown, Boston.
17. McWhinney, I. R.: 1975, 'Family Medicine in Perspective', *New England Journal of Medicine* **193**, 176–179.
18. Myers, W. E.: 1978, 'Parenthood or Politics', in J. P. Brady and K. Brodie (eds.), *Controversy in Psychiatry*, W. B. Saunders, Philadelphia, pp. 801–812.
19. Nagel, T.: 1980, 'Sexual Perversion', in A. Soble (ed.), *Philosophy of Sex*, Littlefield, Adams, Totowa, pp. 76–88.
20. Rubin, I.: 1972, 'Concepts of Sexual Abnormality and Perversion in Marriage Counseling', in H. L. Silverman (ed.), *Marital Therapy*, Charles C. Thomas, Springfield, pp. 435–441.
21. Shelp, E. E.: 1982, 'To Benefit and Respect Persons: A Challenge for Beneficence in Health Care', in E. E. Shelp (ed.), *Beneficence and Health Care*, D. Reidel Publishing Co., Dordrecht, pp. 199–222.
22. Siefkes, J. A.: 1975, *Siecus Report* 3/3 (January).
23. Soble, A.: 1980, 'Masturbation', *Pacific Philosophical Quarterly* **61**, 233–244.
24. Solomon, R.: 1980, 'Sexual Paradigms', in A. Soble (ed.), *Philosophy of Sex*, Littlefield, Adams, Totowa, pp. 89–98.
25. Sommerville, M. A.: 1982, 'Birth Technology, Parenting and "Deviance" ', *International Journal of Law and Psychiatry* **5**, 123–153.
26. Steinert, Y. and Levitt, R.: 1978, 'The Teaching of Human Sexuality in Family Medicine', *The Journal of Family Practice* **7** (5), 993–997.

J. ROBERT MEYNERS

SEX RESEARCH AND THERAPY: ON THE MORALITY OF THE METHODS, PRACTICES AND PROCEDURES

The emergence of modern sex research and sex therapy as an important and recognized discipline has posed a multiplicity of intriguing, even agonizing, moral questions. For such a new and somewhat amorphous discipline, this is certainly not unexpected. However, the intensity of these questions has been the more dramatic because of the special moral and emotional sensibilities surrounding the subject of sex. The cries of moral outrage elicited first by the Kinsey research ([6], [5]) and later by the work of Masters and Johnson [19] point both to the importance of sex in human life and to a popular reluctance to submit human sexuality to scientific scrutiny or intervention.

The following discussion of the moral issues in sex therapy and research will first address the social and historical context in which modern sexual science has emerged, considering the factors that have significantly impacted on moral deliberation in this field. Second, a methodology for considering moral questions in sexual science will be suggested. Finally, on the basis of such a methodology, some specific examples of the moral dilemmas encountered in the field will be discussed.

SOCIAL AND HISTORICAL CONTEXT

During the past forty years of historic change in the field of sexual science, a variety of concurrent developments has impacted significantly on the ethical questions related to this new field. First, a major change in the moral sensitivities of scientists resulted from the shock of Hiroshima and Nuremburg. If it were possible prior to these cataclysms for most scientists to carry on their research without giving much thought to the impact of the work on human destiny, there were few who claimed that privilege thereafter. Specialized journals and books now address

Earl E. Shelp (ed.), Sexuality and Medicine, Vol. II, 171–195.

the ethical implications of scientific research, including both popular and professional discussion. Some examples are *The Bulletin of the Atomic Scientists, The Bulletin of Science, Technology and Society, Science Technology and Human Values, The Journal of Medicine and Philosophy, The Journal of Medical Ethics, The Hastings Center Report*, etc.

Second, this has been a period of new and increased legal and governmental attention to the protection of clinical and research populations. This is a part of the larger phenomenon, a radical proliferation of governmental regulation in general. Among the direct results of this development have been legislative investigations of the rights of human subjects in research. Legal requirements describing what is and is not permissible in such research have been delineated and institutional procedures for the protection of human subjects have been specified.

Third, these years have seen a veritable explosion of biomedical research and technology, posing an astonishing number of new moral dilemmas. Research in many subjects, from isolation of carcinogens to genetic investigations, has had wide-ranging social, political, and economic implications. The resulting technologies have produced scarce commodities, from transplant organs to life-support systems, setting up difficult choices about who gets what and under what circumstances. Technical and ethical questions are inextricably intertwined, as for example in seeking to define death under these new circumstances. Attention to these dilemmas has attracted attention to and shed light on a whole range of moral issues shared by sexual science with other biomedical fields. In addition, these new technologies have posed dilemmas of a specifically sexual nature, as in contraception, abortion, and *in vitro* fertilization.

University departments, courses in medical ethics, organizations and conferences have responded to these developments. Among the most important responses have been the founding and impressive achievements of the Hastings Center, Institute of Society, Ethics and the Life Sciences. Founded in 1969, it has succeeded in bringing together scientists, attorneys, politicians, philosophers, and theologians in an ongoing debate intended to hold both the scientist and the institutions of public order to moral accountability. Laboratories, journals, and factories are searched for scientific and technological developments needing ethical consideration. Classical ethical traditions and peculiarly modern moral perspectives are brought to bear on these new problems. Scholars are provided with time to study and reflect through appointments as fellows, *The Hastings Report* is published to air the general ethical issues in biomedical research, while *IRB: A Review of Human Subjects Research* is addressed to the theory and mechanics of research accountability, especially through review boards. Two series of volumes have been addressed to ethics and ethical foundations in relation to science. Sexual science has benefited from these achievements because of the essential clarification of the ethical issues, but it has also benefited in direct ways when the subjects analyzed were specifically sexual [10].

In the first burgeoning decades of the new field of sex therapy and research, scant professional attention was directed to the ethical issues. Public discussion of the moral quandaries the scientist faced would no doubt have provided support for those who regarded the entire enterprise as immoral. Certainly individual scientists and clinicians were privately concerned with ethics, but professional attention was not publicly addressed to these problems until the mid-seventies. Prior to that, there were three kinds of restraints that functioned in the field.

First, there existed a myriad of laws, federal, state, and local, which addressed sexual behavior and which therefore could impact on professional decisions. Knowledge of these might help to keep the scientist or clinician out of court, but they were of little assistance in providing guidance for more profound questions. A more serious weakness was, and is, that there exist many archaic laws, whether enforced or unenforced, representing values no longer generally honored in our society. Thus, ethical responsibility on the part of the professional was influenced by this chaos, and legitimate professional endeavors were often in legal danger.

Second, social mores provided significant restraints on the work of sexual scientists and clinicians. However much it may have been violated in private, American society has given homage to a relatively narrow range of permissible sexual behaviors, even by comparison to Western European nations. With Puritan roots and frontier pressures, Victorian influences and a commercial ethic based on public propriety, American history exhibits a highly privatized understanding of sexuality, which is, to say the least, dissonant with exposing sex to scientific observation or clinical intervention. This state of affairs has exercised a restraint on the activity of scientists in this field, which was directly or indirectly based on moral considerations. Moreover, the decades under discussion are marked by important changes in social mores with regard to sex, changes so rapid they have been dubbed "the sexual revolution." Sexologists may indeed have influenced these changes, but in any case, these developments introduce a new degree of ambiguity and controversy into the patterns of moral expectation.

Third, there were codes of ethics in the various disciplines from which sexual scientists and clinicians approached their endeavors. Medicine, social work, and psychology each had professional codes. Pastoral care, social work, and sociology had their codes and ethical traditions. These provided an excellent basis for ethical decision for the members of these professions, but they did not provide for any special questions that the subject of sex might pose. During the decade of the seventies, professionals in the relatively new field of marriage and family therapy were increasingly concerned with the special problems posed by the presence of more than one client in therapy, a situation shared by conjoint sex therapy.

Since sex research and therapy was a new interdisciplinary field, people entered the field who did not belong to any of the traditional health care professions and, therefore, did not have such a code of ethics as a part of professional tradition. The problem of clinical credibility became urgent, since virtually anyone could practice sex therapy. Individuals with excellent training in their original field could and often did practice sex therapy with little or no preparation. Others entered clinical practice with no previous training or supervision in psychotherapy.

This was the situation a decade ago when ethics in sex therapy and sex research surfaced as a significant professional problem. Out of this concern, the Masters & Johnson Institute (then the Reproductive Biology Research Foundation) sponsored two conferences addressed to this subject. At the first conference held in January, 1976, a group of nationally recognized specialists in the field were joined by lawyers, theologians, and philosophers, identifying and addressing the ethical issues raised by sex research and therapy in its developing forms. Scientific, historical, theological, and philosophical perspectives were outlined, and the differences in ethical problems between research and therapy were delineated. The problems that emerged as most critical were those of confidentiality, the rights of privacy, informed consent and standards of clinical competence. The published proceedings of the conference [14] formed the basis of general discussion in the field.

A task force was then recruited to prepare background papers and a set of ethics guidelines, which were to be considered at a subsequent Ethics Congress. A somewhat larger group was convened in January, 1978, and the proceedings were published in a second volume [15].

On both these occasions there were significant differences of opinion. There was disagreement on whether there should be ethical guidelines at all, at least at that time. Alternatives suggested were a voluntary review body or an ethical casebook that would provide a developing data bank of information compiling the experience of people in the field. A further general disagreement arose over whether sex research and therapy were substantively different from other fields in biological and psychosocial science. There were also vigorous disagreements between those who believed confidentiality was absolute, without consideration of third party welfare, and those who believed confidentiality to be relative in the case of conflict with other countervailing claims.

The guidelines were revised in the light of the discussion at the second meeting and became the basis of further discussion in the field. In revised form, they were adopted by the American Association of Sex Educators, Counselors and Therapists, and subscribing to them is one of the requirements of accreditation by that body. Currently, a further revision of the Guidelines is under consideration by AASECT, because of both a desire to shorten the document and to remove the standards of training and competence, since these are covered by accreditation requirements.

A METHODOLOGY FOR MORAL DELIBERATION IN SEXUAL SCIENCE

Before we can address directly questions of the morality of sexological procedures, we must make some choices about the methodology for doing so. There are several alternative ways of approaching the matter. First, we can take a specific moral methodology and use it to examine the questions. Thus, one could use a religious tradition and its values as the measure, or one could employ a specific philosophical methodology such as positivism or idealism. The richness of these traditions produces meaningful discourse, but the conclusions are satisfying only to those who share similar assumptions. Such a debate also shifts the focus of disagreement to questions that have nothing specifically to do with the subject of sex. Since the wisest minds over countless centuries have occupied themselves with these basic questions, and without agreement, we cannot wait for their resolution to decide for example, what to do now about observation in sex research or the use of surrogates in sex therapy.

A second approach would be to recognize that no agreement is possible on moral principles and to content ourselves with admonishing everyone to be ethically concerned, depending on each individual to make responsible decisions in the field. The argument is: "If history has not provided us with a consensus, how can we in a pluralistic society hope to achieve agreement, either on moral principles or on moral conclusions." Such individualism and relativism is indeed the temptation of a democratic society, but it removes the individual from accountability while it weakens the possibility of significant agreement and trivializes disagreements.

A third approach is to assess the social fabric to determine what the public attitude is about the moral questions involved in sexual science and to make this the measure of accountability. The two ethics conferences referred to above give ample evidence that such deliberations must take account of public opinion. Indeed, no serious discussion of the morality of sexological procedures is likely to be uninfluenced or unconcerned with social mores on the subject. This fact has important implications for an understanding of professional ethics and the moral dilemmas posed by professional activity. Radical departures from social norms create hazards for research and clinical practice, the more so in a socially sensitive field like sex. On the other hand, uncritical adherence to these norms may be a disadvantage to scientific discovery and clinical effectiveness.

What then is the role of social mores in the professional ethics of sexual science? A poll of what the public thinks about measuring the physiological response of orgasm would hardly provide an adequate basis for deciding whether or not this procedure should be employed. Nor will a sociological poll ever by itself tell us what is morally right. However, what is useful about a moral approach that is based on social mores is that it starts from what cannot in any case be avoided, namely, the human and historical context in which the decisions are made. No moral approach is feasible in an ordered society that is not based in some way on what most people regard as good, even if that definition must be submitted to further critical reflection. No ethical code for sexual scientists could be based on the assumption that lying is good, in a society that honors truth–telling both in compliance and breech.

This, however, poses a further problem. If most individuals make their moral judgments by habit or intuition, they will not be able readily to define what their values are. It is a fair guess that most people do not reflect on the values lying behind their moral predilections. Nevertheless, moral judgments in a scientific field are sooner or later in the public domain, and the decisions scientists make will impact on a public whose reactions will affect the consequences of the decision. While there may be a measurable popular opinion on moral questions related to sex research and therapy, there is not and never will be a public consensus on either the conclusions or the values that lie behind those conclusions. Thus, a consideration of social mores is a necessary but not sufficient desideratum of moral norms in the ethics of sexual science.

Social rules ordinarily posit conclusions without consideration of the basis of decision-making or the principles behind the conclusions. Thus, each new situation is approached as if it were an old situation. In previous centuries the meaning of a male touching or observing a female body had only sexual meaning, or occasionally an artistic import. In the meantime, a variety of other meanings is possible. Consider the example of a physician palpating the female breast. If the action is considered without regard to its purpose (disease detection), technical effectiveness (x percent reduction in cancer deaths) and social safeguards (physician's office, chaperone, etc.), it would be considered morally wrong unless of course some other set of circumstances made the action sexually permissible (e.g., they were married). Thus, in the search for ethical guidance in sexual science, the rules of ordinary human intercourse cannot be simply translated into professional rules. It is for this reason that the question of the morality of the procedures used in sex therapy and research are in the present essay considered in the context of a professional ethic, rather than as moral questions independent of the scientific context.

However, a professional ethic for sexual science must have a certain universal character. For example, the stipulations must not be dependent on the individual scientist acting out of altruism – since obviously not all scientists will feel it their duty to act altruistically. Certainly some scientists will be choosing actions on the basis of their own advantage. Note, however, that such an individual may be making exactly the same behavioral choice, although on a different principle. One clinician may choose not to have sex with a patient out of concern for the welfare of the patient, while another clinician may choose not to have sex with a patient because of fear of punishment. As long as there are rules and effective sanctions to enforce the rules, the resulting behaviors will tend to be the same or similar.

Whether Hobbes was correct or not, that everyone always acts from self-interest, it is at least clear that some of us some of the time act from self-interest, and there is no reason to believe that scientists are significantly different from the rest of human nature in this regard. There are two possible responses to this state of affairs. One is the idealistic response of a Joseph Butler who assumed that, if people acted truly and rationally from self-interest, they would do the same thing as if they acted from altruism. But there is no more reason to assume that people will act with consistent reason than to assume that they will act with consistent altruism. The other response is that of a social contract theory in which social concourse is based on an implicit or explicit agreement to organize society in some way serving the good of all (Hobbes, Locke, Rousseau, etc.). It may seem like a long jump from the social contract theories of the 17th and 18th centuries to a modern professional ethic for sexual scientists, but a professional ethic is an implied contract between members of a profession and between the profession and the larger society. Whatever the values are that individuals hold, a rule of law is required in order for them to be free and able to pursue those values. Whatever values the individual members of a profession hold, a convention for professional behavior is required if they are to be free and able to pursue those values.

These deliberations support Peter Caws when he delineates the following general propositions: that some form of moral rule becomes a value in common for people required to live in a common world; that the moral rules are held for the sake of values about which there is no agreement; that the rules are invoked in response to a consideration of what might happen without them to the private aspirations of a lot of free individuals inhabiting the same planet, and that the general import of the rules can then be derived from a consideration of the various ways in which such aspirations might be frustrated [4].

The question then becomes what are the principal ways in which human aspirations might be frustrated. Caws suggests that there are three: the individual might be deprived, might be deceived, or might be interfered with. For our purposes in developing a professional ethic, the first two may be subsumed under the third. The professional is not in a position to judge what deprivation would mean to the individual, since he or she is not an expert in any other area of human existence. A major hazard of professionalism is to assume a paternalistic attitude toward clients. Technical competence may provide some basis for such judgments, but the individual is ultimately responsible for the values he or she wishes to protect or pursue both within and outside of the therapeutic context. If the client or research subject is deprived, he or she will also be interfered with; if deceived, then also interfered with. There is therefore a logical priority to the criterion of non-interference. However, there is an empirical basis for this priority, since pluralism is a result of freedom. Not only in democratic society, where multiplicity of values are ever-present, but even in more authoritarian societies, whatever pluralism exists is the product of some degree of freedom. If a completely totalitarian society existed, there would be no variation in the values individuals hold. Non-interference is freedom made operational.

A professional ethic is above all operational; it is not a theoretical moral system. A general moral system would have to describe a concept of justice or fairness by which deprivation is to be recognized and a concept of truth by which deception is to be recognized. Rather, a professional ethic needs a system of rules by which professional behavior is regulated, so that (1) persons who need and wish therapeutic services are not interfered with in pursuing their own values, (2) so that the professionals providing these services are not interfered with in pursuing the values they hold in providing such services, and finally, (3) so that the public, in so far as the public is affected by these transactions, is not interfered with in pursuing whatever values are present in a popular consensus.

The values represented by these three participants in sexual science may be competitive and thereby exists the need for rules or guidelines to regulate this competition. Such deliberations are essentially prudential, even pragmatic. They are designed to provide for the participants freedom to pursue their own values without unnecessary interference. The scientist is not responsible for protecting the values of the client, subject, or the public, but only for not interfering with their pursuit of values and for balancing the claims that are made on his or her resources.

The professional ethic establishes a convention. It is not surprising, therefore, that, prior to 1847, when the American Medical Association defined a code of ethics, medical ethics was commonly referred to as medical etiquette. A charge frequently brought against the adopted codes of various professions is that they do not address morality but etiquette. If they do in fact do so, that may indeed be a mark of their success. Another charge is that these codes protect only the profession and not the public, but of course if they in reality protect the profession, they can only do so in general consonance with the public good. The specific actions of an individual may profit the individual to the detriment of the society, but if the codified guidelines for the conduct of a profession are detrimental to the public, the profession will in general suffer. Basically, a code of professional ethics seeks to bring into consonance the self-interest of all affected by the professional activity, just as etiquette at the dinner table dictates that one act in a way that does not upset the occasion for the other participants.

Our conclusion so far is that the morality of the procedures involved in sex research and therapy is to be measured by rules or guidelines that adjudicate whatever conflicting interests are involved between the various parties affected by the procedure. Such rules are to be based on the dictum that no party affected shall be interfered with in its pursuit of its cherished values, and the rules are to provide guidelines and procedures to the professional in balancing any conflicting claims. The interested parties are the patient or research subject, the public, and the scientist. This approach provides latitude for the wide range of values held by the people affected by sex research and therapy, as well as for the values held by the professionals who conduct these activities. Moral responsibility can be exercised in the highly pluralistic environment in which sexual science exists.

This leads to one final and critical point. It will hardly have escaped anyone's attention that the scientist does not necessarily hold the same values as the subject or patient affected by his actions. However, since the approach we are suggesting does not base on the commonality of values held among the participants, but only on a system of rules that enables each to pursue the values held variously and individually, it follows that the situation is quite amenable to moral responsibility so long as the scientist respects, in theory and in practice, the values of those affected by his professional activity.

We turn now to several topics that illustrate the operation of this approach, beginning with two subjects that have long been matters of significant ethical concern to sexual science. Thereafter we will turn to more novel subjects under current question.

INFORMED CONSENT

No general principle is more central to the discussion of the ethics of sex research than that of informed consent. In this as in other fields of research on human subjects, there is general agreement that the subject should be advised of the nature and purpose of the research, the advantages, if any, to the subject, of participation in the project, the nature and probability of risk, the availability of treatment for any harm sustained. The protocol and procedures must reasonably assure that the participant is not subject to coercion to cooperate and is free to terminate participation at any time with or without cause and without loss of treatment or other pre-existent advantages. However difficult these conditions may be to achieve in practice, their necessity is generally agreed. It does, however, severely restrict the kind of research that can be done. It may make impossible research using placebos, studies that depend on a group of naive subjects, or studies which involve a delayed or non-treatment group.

With regard to therapy, there is a developing belief that informed consent is equally applicable to the therapeutic relationship. Formerly, the general assumption in psychotherapy was that the patient knows what is necessary to enter into an agreement of implied consent. Now, however, some therapists use carefully wrought descriptions of the therapy format, a statement of risks and outcome probability, and a consent form to be signed by the patient. The remaining question is whether such specificity is prejudicial to a positive outcome to therapy.

If patients or subjects are fully and accurately informed about the procedures to be employed and no coercion is present, then they have a choice about whether participation in therapy or research is consonant with values they hold. Therefore, the procedures for providing informed consent are necessary and desirable guarantors of the morality of the interventions to be employed. With some notable exceptions, sex research and therapy present no unique problems with regard to informed consent. Let us examine some of these exceptions, using primarily the criterion of non-interference.

There is disagreement in sex therapy about whether and under what conditions physical contact between the patient and therapist is ever justified. There is general agreement in the field that therapist-patient sex is proscribed, although what this means remains somewhat vague. Heterosexual intercourse is certainly prohibited, but what is the mark of a physical contact that is sexual? Ideally, it would refer to any contact that gives the therapist sexual gratification, but that is so subtle and unpredictable a response that it is useless as a desideratum. A therapist can become aroused in a talk session in sex therapy and that is neither preventable nor particularly harmful unless the therapist's objectivitv is thereby impaired. Conversely, a physician touching the genitals of a patient can be, and probably usually is, without any sexual connotations. Therefore, therapist's arousal or actual physical contact with the patient is neither by itself an adequate criterion for the appropriateness of physical contact between patient and therapist.

A recent suit charged a gynecologist with malfeasance because he stimulated a patient's clitoris, ostensibly in order to determine if the blood supply to the labia was sufficient for engorgement as a mark of sexual excitability. In making a decision about whether to carry out such a procedure, the physician might decide to do so even if it is alien to the public conscience and is not accepted practice in the profession, on the basis of a belief that the procedure is likely to achieve its intended purpose, that the patient will not be harmed, and that not to employ the procedure is to interfere with the patient in her pursuit of sexual health.

However, this case actually illustrates the need for standards of competence as a part of professional ethics. The physician's judgment was impaired perhaps by his own personal needs or by an understanding of sexual physiology that ignored psychosocial factors. If the woman did not respond with engorgement, there would be no way of knowing whether she was physiologically impaired or simply put off by the circumstances or the physician. While the procedure would be judged unethical by the usual stipulation that there be no sexual contact with a patient, the ineffectiveness of such a diagnostic procedure would in any case render it unacceptable. Thus, there is a convincing argument that professional codes of ethics need to state standards of competence and accreditation in order to indemnify the adequacy of technical judgments as an element in moral deliberation. The patient is interfered with if the therapist is not competent to provide the patient with informed consent about the probable consequences of a procedure.

In a second illustration, a clinician has a patient whose presenting complaint is severe angulation of the penis with engorgement and consequent interference with sexual function. The patient's self-report can be checked with a partner if one is available for interview, but even then it might not be possible to distinguish anxieties or hostilities secondary to a mild curvature from a physical condition of sufficient severity to interfere with intercourse. The obvious solution would be to observe the patient under conditions of sexual excitement. To do so, however, might involve professional risk, since the observation probably could not be proven to be accepted practice, and in fact might occasion public outrage if, for example, the partner filed a complaint. There is in this case a conflict between the professional's desire for self-protection and the patient's desire for an informed diagnosis. Further, the patient may place a countervailing value on sexual privacy and a consequent unwillingness to be embarrassed by being observed in a state of sexual excitement. These considerations take account of each person's right not to be unnecessarily interfered with in the pursuit of values each holds. A variety of solutions suggest themselves. A chaperoning nurse or colleague might provide protection for the clinician but also might add to the discomfort of the patient, thus exacerbating the already present risk of iatrogenic trauma. Or the techniques of artificial erection might be applied, but these are neither commonly available nor accepted practice, and might be questioned as an unnecessarily invasive procedure under the circumstances. There is finally no easy solution to the physician's dilemma.

These examples illustrate that moral values commonly held in society and the profession can neither be ignored nor slavishly followed in an adequate professional ethic. Reflection on the moral values current in the society produces a limiting influence on professional behavior. Reflection on the consensus within the profession places a further limitation on professional procedures. Finally, the personal values of the professional may pose a still further limitation on the latitude of choice. The logic of both these illustrations is that moral reflection on scientific and clinical procedures can be expected to produce conservative and limiting influences over the range of professional choice. This is perhaps an important reason why sexual science was so long in emerging. The difficulties that Kinsey and Masters and Johnson encountered in developing their work read like a bad Victorian novel, and we have learned that a simple use of public conscience or professional custom are an insufficient basis for responsible decision.

The primary moral imperative suggested here is the right not to be interfered with in the pursuit of values. If the patient is not fully informed or is deceived, he or she experiences interference. The patient, the professional, and the public are deceived by any failure to determine effective procedures and to disseminate information about their effectiveness. This provides the basis of placing value on innovation as one of the features of an adequate professional ethic, and justifies calculated risks when the possibly conflicting interests of all concerned are carefully weighed.

There are two problems that, if not unique to sex research and therapy, are at least especially difficult in this field. The first has to do with the values held by the researcher or clinician and the possible right the subject or patient may have to know of these values. A cancer researcher may have a deep concern in the field that, for example, is intensified by the loss of a loved one to the disease. However, this will be unlikely to influence negatively the treatment of a patient or subject. By contrast, a sex researcher or therapist may be profoundly influenced in the treatment of a sex offender if he or she has been the victim of such an offense. While this problem of value bias is shared by other kinds of behavioral research and therapy, there are some issues that are even more specific to sex.

One example of this is occasioned by a homosexual subject or client with a heterosexual researcher or clinician. No matter how much the scientist may believe himself/herself free of prejudice, social conditioning is unlikely to leave any heterosexual untouched by a value bias. Value–free therapy is a malicious myth [20]. It may not be much help to arrange that homosexuals be treated or studied by homosexuals, since that is also no guarantee of a value–free perspective. It may be useful for the enterprise to be carried on by a mixed team, but this will not always be practical. Some scientists deal with this problem by providing a statement of sexual orientation and providing an opportunity for the client to ask exploratory questions. This may enable the prospective subject or client to assay the quality of value bias of the professional to whom his or her welfare is entrusted. However, no such procedure relieves the professional of the responsibility to seek to behave in a way uninfluenced by the difference in value perspective.

A second example is occasioned by the gender of the researcher or therapist ([12] [7] [9] [17]). No value bias is more invisible to the holder than those unique perspectives that are the result of biological differences and social conditioning with regard to gender and gender roles. A task force of the American Psychological Association found family therapists particularly vulnerable to biases (1) assuming that remaining in marriage would result in better adjustment for the woman,

(2) showing less interest in the woman's career than the man's, (3) assuming that child responsibilities were more the woman's than the man's, (4) manifesting a double standard with reference to sexual affairs engaged in by the female as compared to the male, and (5) deferring to male needs [1].

These issues are centrally involved in the dynamic of sex therapy, and the difficulty cuts both ways. A couple may present in therapy with traditional values about male dominance in the relationship. The therapist has three choices: to honor the couple's values and ignore the negative effect those values may have on therapeutic outcome; to instigate therapeutic procedures based on a different understanding of marital relationships and which ignore the couple's value system; or to explain the therapeutic bias, and allow the couple's informed choice to prevail. The latter choice best conforms to the conditions of moral choice in professional behavior, on the basis of non-interference. If the latter choice is followed sensitively, informed consent obtains. However, this may be too simple since (a) the couple may disagree, (b) their values may change in the course of therapy, even for extraneous reasons, or (c) the therapist may share the value system of the couple, thus hiding the disadvantages of this value orientation. The far more serious problem, then, is the male bias that may enter the therapeutic situation, sometimes at the hand of a female therapist. A female therapist may reasonably be expected to reduce this bias, but the admonition that no one in the therapeutic situation should be deprived is the goal of the rules surrounding informed consent. Again, value-free therapy is an ideal, but not an achievement. Since it is only a fragile possibility, there must be both principles and institutional procedures to minimize the risks of value bias to the female client in conjoint therapy.

CONFIDENTIALITY

Rigorous concern for confidentiality is a part of professional ethics in sex therapy and sex research as in other clinical and research fields, and for the most part there is little that is unique to this situation. In research there are difficult problems, since there are so many research methodologies that are only effective if confidentiality is absolute, and yet absolute confidentiality may be difficult or impossible because of legal requirements or expense. In therapy, problems of confidentiality are compounded by legal ambiguities about third party communication [13], by requirements to report venereal disease, and mandatory reports when there is suspicion of child-abuse, or reason to believe that a felony may be committed. These issues lap over into informed consent. If, for example, a father seeks therapy because of having molested his daughter without knowing that the therapist is required to report to a government hotline, he has not been given the right of informed consent. However, if he is told of this requirement, he may be influenced to withhold necessary information or to decline therapy and both the victim and the public may thus be subject to interference. These are difficult ethical issues because there are conflicting principles and competing interests involved. They are difficult moral issues because it takes time, effort, and even money to provide the procedures that best indemnify the ethical principles employed. Of course, these problems are paralleled in any form of psychotherapy.

The fact that sex is so sensitive a subject for many people may heighten the damage done to them when there is any breach of confidentiality. Thus there is a need for rules and procedures to protect the individual from either trivial or serious harm. Even the information that one has been in sex therapy is confidential information, which may be released only on the basis of signed consent or legal requirement. There is a curious irony in that government agencies charged with the protection of human subjects can require access to the names of research subjects in order to determine whether confidentiality, informed consent, and subject safety are adequately protected.

There are special problems of confidentiality involved in couples' therapy and research done with couples, even though these are not unique to sexual science. What is to be done with information told in confidence to a therapist with the expectation that the information is not to be shared with the partner? There are three characteristic ways of handling this: (a) an understanding at the outset that there is no secrecy "so don't tell me anything you do not want shared", (b) an understanding that anything told by one member of the dyad to the therapist in private is and will remain confidential from the other, (c) an understanding that either has the right to request that certain information be made confidential [13]. None of these is very satisfactory. The first disadvantages the therapy and therefore deprives the couple, the second and third may maintain one or both clients in ignorance and be disadvantageous.

A variation on these approaches is to have an initial understanding that information designated in private as confidential will remain so unless and until the therapist decides that the material should be dealt with in therapy, at which point the client is told in private the reasons and given the choice. This has the disadvantage of putting an inordinate responsibility on the therapist, and in any case may not deal with difficult cases, but the approach is highly workable in practice if the therapist is sensitive and skillful. When the issues get complicated, the moral responsibilities of professionals require time, effort, skill, and even money in order that competing principles and interests may be satisfactorily adjudicated.

THE USE OF SEXUAL SURROGATES

When Kinsey published his work, there were two popular moral concerns. First, that people should be asked such personal questions violated the current public norm that sex is a private matter that should not be invaded by scientific interrogation. Second, the published results seemed to indicate that behavior violating social mores was widespread, and therefore many people, especially young people, might be led to believe that such behavior was morally acceptable. When Masters and Johnson published their work, there were again two common concerns in the public discussions. First, that people should be observed and measured in sexual activity again violated the current public norm that sex is a private matter and should not be invaded by scientific investigation. Second, that these investigations might reduce sex to a biological process not subject to moral judgments or considerations of social propriety engendered a fear of hysteric proportions. While these concerns continue today, the voices of alarm are muted by the sounds of social change, whether these sounds are regarded as noise or music. The concept of consenting adults has replaced an earlier iron-clad understanding that the public has an interest in protecting privacy by requiring it. In few jurisdictions is non-marital intercourse subject to legal sanction.

While the concept of "consenting adults" has provided an increased moral freedom for the sex researcher and the sex therapist, to say nothing of individual partners and lovers, it has by no means provided an ethical panacea. The consenting adult, whether professional or personal, is still faced with moral questions about the health and usefulness of alternative behaviors. Assault is still illegal, and no doubt immoral, and the fact that a partner consents in sado-masochistic passion to mutilation (whatever consent might mean under these circumstances) is not without a public investment. Nor is a researcher studying the behavior of a couple engaged in such activity relieved of moral responsibility to consider any public investment in this private behavior. Guidelines and procedures for the protection of human subjects are formulated not only to protect the individuals involved, but also to protect the public from a scientific ethos that intends to be value–neutral while it projects an amoral perspective on problems of profound significance for human well-being.

The moral issues faced today in sex therapy and sex research are both more subtle and more sophisticated than those that greeted Kinsey or Masters and Johnson. The medical model of professional ethics and the conventions of sociological and anthropological investigation had largely disposed of the gross and superficial questions of professional behavior in the presence of nude bodies, or palpation of sex organs for diagnostic purposes, and even to some extent the asking of personal questions covering sex and intimacy. Appropriate if inadequate safeguards had been developed in terms of professional codes and traditions. While the next decades were to refine these safeguards, a process that continues uncompleted, it was actually the size, scope, and credibility of these projects that made them so threatening to conventional morality. We now must wrestle with fine-tuning the professional guidelines and procedures that best protect the individuals and the public from interference.

In order to illustrate this new situation concretely, it will be useful to consider the morality of the use of sex surrogates. Sex surrogates are professional or paraprofessional persons who act as substitute sexual partners for individuals entering sex therapy without a partner of their own. It is interesting that in the reactions to the initial publication of the work of Masters and Johnson, the use of surrogate partners in therapy was not a special focus of moral outrage. This probably represents a certain "folk wisdom" that the moral implications of this work were far more profound than any charges of turpitude might imply.

The perspective on professional ethics that we have been outlining presumes that the therapist has a responsibility to treat people when they request therapy and when there is reasonable belief that the therapist has the competence and resources for effective treatment. There is a public trust that goes with the professional role. This is a vocational responsibility defined by the support the public provides in the freedom to pursue the profession. The patient and the public are interfered with if the professional arbitrarily decides not to treat or study an appropriate patient or subject.

The right not to be interfered with, as we have argued, is an operational way of conceiving the responsibility for the welfare or good of those immediately affected by one's professional activities. Individuals or couples suffering from sexual dysfunction accordingly are interfered with, if therapeutic services are denied them. The complications of medical and therapeutic economics do not require that every presenting patient be treated regardless of whether the service can afford it, but sex therapy shares with other health care professions very difficult responsibilities to provide care to those who cannot afford it, consistent with the capacity to earn a living and continue to provide the service at all. However, the economics of sex therapy is too large a subject to be dealt with here, even though there are important ethical implications and dilemmas.

If there is a prejudice in behalf of providing therapy for those who require it, what then is to be done when an individual presents in therapy without a partner? The choices seem to be: to deny sex therapy to the individual, to refer the individual to a form of therapy designed to overcome the obstacles to forming intimate relationships, or to provide sex therapy with the use of a sex surrogate.

There are difficulties with each of these choices. There are often legitimate reasons for a therapist to refuse treatment, and if the therapist has reason to believe that surrogate therapy would be an escape from the responsibilities of intimacy, or if there is an alternative form of therapy that holds greater promise, then the question of using a surrogate is moot. However, if the therapist rejects surrogate therapy because of personal moral concerns that the client does not share, then that refusal is open to question, since interference with the client's pursuit of the values of sexual health are subject to interference. This might amount to imposing the therapist's values on the client. On the other hand, if the therapist provides surrogate therapy (a) without considering other alternatives, (b) without having reason to believe that such therapy has a probability of positive outcome, and (c) without considering the possible disruption to the value system of the client, then serious questions might be raised about this procedure.

It is clear, then, that a responsible moral judgment about the use of sex surrogates involves a technical judgment about effectiveness. Unfortunately, there is no consensus on this within the field. Meetings of the professional societies sometimes hear papers read for and against the use of surrogates. In the small attention given to the subject of surrogate therapy in the literature of sex research, there is a similar division of opinion. Some have reported positively on therapy with surrogate partners ([16] [18]). On the other hand, Kaplan takes the position that either the neurotic difficulties that have prevented the person from developing relationships can and should be first dealt with before undertaking sex therapy, or a form of sex therapy that does not depend on the availability of a cooperative partner, such as psychoanalysis, should be employed [11]. Apfelbaum, while approving of the precautions that Masters and Johnson employed, argues that surrogate therapy is a myth because it creates performance anxiety in the surrogate and thus is likely to reinforce performance anxiety in the patient [2]. Making a judgment about the effectiveness of surrogate therapy is complicated by the absence of outcome studies, a weakness of sex therapy in general, and in fact of psychotherapy in general. Similarly, Helen Kaplan's faith in psychoanalysis for the treatment of sexual problems, even if true, is not as yet a scientifically demonstrated hypothesis.

This is not the place for a definitive discussion of the effectiveness of surrogate therapy, much less the inherent difficulties of effective outcome studies, but rather to point to the moral dilemma occasioned by the lack of a professional consensus on the subject. The apparent paradox is that many, perhaps most, sex therapists do not employ surrogate therapy, because it is not yet commonly accepted practice, and those therapists who do use the approach are hesitant to discuss it or write about it openly. Under these circumstances, the relevant clinical experience and empirical investigations are difficult to achieve. Thus the questionability of the procedure prevents the development of information necessary for an informed technical or moral judgment about the acceptability of the procedure.

A second way in which technical questions have ethical implications in the use of surrogates relates to the therapist's responsibility for the competence of the surrogate. There are different understandings of the surrogate role. Is the role to be a sex expert and to some extent a therapist, or alternatively to be a partner who can aid the therapy because of basic comfort with, and information about, sexuality? In either case, responsibility for the case places responsibility for the surrogate on the therapist. Whether the therapist provides training or utilizes the resources of a professional organization of surrogates, criteria of selection and appropriate handling of the body-work situations must be applied.

An extension of this same issue is the therapist's responsibility for the welfare of the surrogate. Again, there are role ambiguities. Is the therapist the employer of the surrogate with the same responsibilities as any employer to an employee? Is the surrogate an outside contractor where the therapist has more limited responsibilities? Is the surrogate the employee of the client, where the therapist has even more limited responsibilities? Whichever of these understandings is chosen, the actions of the therapist impact directly on the surrogate, and he or she becomes one of the parties affected by the therapeutic situation. If a therapist chooses to involve surrogates in therapy, the surrogate becomes one of the parties deserving of consideration in any conflict of competing values.

RESPONSIBILITY IN SOCIAL CHANGE

There can be little doubt that sexual science has influenced and been influenced by changes in the social fabric over the past forty years. Attitudes about many sexual subjects have evolved significantly. Premarital sex is generally accepted and practiced, while extramarital sex is no longer the subject of unmitigated outrage. Masturbation is commonly regarded as healthy, and moral strictures against it are in constant retreat. Homosexuality is more and more considered as simply an alternative sexual style. Transvestites, transsexuals, and non-injury paraphiliacs are treated for any ego-dystonic aspects of their propensities and otherwise may find supportive treatments in pursuing their sexual goals. This list of private sexual behaviors that are either condoned, accepted, or tolerated could be expanded indefinitely. Sexual science, whether intentionally or unintentionally, has participated in these changes.

This fact carries with it moral responsibility. This statement does not imply that science should refrain from activities that may influence social change, nor that science should actively propagate social change. Rather, it is to state unequivocally that one of the moral responsibilities of the sexual scientist is to reflect seriously about the impact of scientific activities on the social fabric and the values woven into it.

A few years ago, K. G. Bailey wrote what could only be interpreted as a shrill and moralistic critique of a treatment program for sexual dysfunction, which included masturbation. Not surprisingly, the article occasioned immediate and forceful peer reaction. However, there was an important concern behind the misguided moralism that was ignored in the legitimate and appropriate criticism that followed. Bailey enjoins a professional ethic to guard against the abuse of the power to direct and facilitate social change. "Just as Freud, Skinner and even Spock have effected massive social change directly with new ideas and techniques, and indirectly with numerous covert stipulations that emanate from their work, the modern sex therapist carries a powerful message to a credulous public looking to psychology for guidance. Although the field of psychology in general implicitly espouses exaggerated individualism, liberal ideology and instant gratification, it is the sex therapist who most seriously assaults the traditional institutions of marriage, family and religious values" [3]. Behind the exaggeration and complaining tone of these words, there lies obscured an important ethical concern. A more restrained view is that of Halleck who believes that all psychiatric intervention has political consequences and that the therapist should evaluate the impact of therapy on the social system as well as his interaction with the patient [8].

In a democratic society we trust the competition of ideas and values to produce the best results, but that relieves no one of the responsibility to think about the social implications of their activities. The timberman cuts down the tree to make houses and fires, to clear land and grow food, but if he is not concerned with reforestation and the prevention of erosion, there may someday be no forest and no food. The analogy is worth pursuing. The love of the magnificent old oak tree may produce a sentimental decision that perpetuates want and prevents progress. The fear of eroding values is not a sufficient basis for restraint in sexual science. Rather, a balanced consideration of older values and newer possibilities may hopefully avoid the pitfalls of thoughtless experimentation on the one hand and blind faith in values, which no longer work to everyone's best interests, on the other hand.

It is probably safe to say that, as a field, sex therapy and sex research are unequivocally on the side of social change. Therapists are aware of the tremendous pain and trauma people experience as a result of restrictive sexual values and the aberrations that are the product of unsuccessful attempts to conform to those restrictions. It is no accident, then, that a conscientious sexual science will often, if not usually, be involved in cultural criticism. However, the scientist can all too easily be over-confident of judgments about the future. A rigorous science will be slow to depart from an empirical methodology that predicts with caution. Recently a father of a young child said, "I don't want to bring up my children with the sexual moralism that I experienced, but I do not know whether the different style we now have in our family will turn out the way we want." He and his wife are making the wisest choice they know how, but there are no generations of experience behind those choices to provide comfort and confidence. Often sexual science sounds as if it knows more than it does about that future.

Nevertheless, it is the responsibility of the scientist to make the best judgments possible and to disseminate that information. Uncertainty there may be, but the sexual scientist is nevertheless in a better position than most to delineate the consequences of value choices on sexual health. The scientist interferes with the public's right to know if timidity or confusion obscure the results of research or clinical experience. The therapist is depriving the patient if the consequences of the values the patient holds are not discussed, but it is still the patient's choice to retain or modify these values. It is a part of the patient's right to know that those consequences be outlined. When the scenario of therapist discussion and patient decision is reenacted thousands and thousands of times, the social impact will be considerable. The emerging judgments, whether probable hypotheses or demonstrated conclusions, will be published and become an important influence in social change. Since pure objectivity is more a goal than a reality, accepting the responsibilities of influencing social change is fraught with danger, but that responsibility is avoided at even greater cost.

In this area, as in other areas of the ethics of sexual science, there are no easy answers. This paper has not tried to provide answers to moral dilemmas, but rather has attempted to suggest and illustrate a framework for examining those questions. These considerations are necessarily vague unless and until they are translated into guidelines and procedures with appropriate sanctions. Thus, the discussion has evolved around the possibility of a professional code of ethics that might embody a relative consensus of professionals in the field. While such a code currently exists, there is a need continually to refine the documents and traditions that hold out the probability of a maturing professional ethic.

Masters and Johnson Institute,
St. Louis, Missouri, U.S.A.

BIBLIOGRAPHY

1. APA Task Force: 1975, 'Report on the Task Force on Sex Bias and Sex-Role Stereotyping in Psychotherapeutic Practice', *American Psychologist* **30**, 1169–1175.
2. Apfelbaum, B.: 1977, 'The Myth of the Surrogate', *Journal of Sex Research* **3**, 238–249.
3. Bailey, K. G.: 1978, 'Psychotherapy or Massage Parlor Technology? Comments on the Zeiss, Rosen and Zeiss Treatment Procedure', *Journal of Consulting and Clinical Psychology* **46**(6), 1502–1506.
4. Caws, P.: 1978, 'On Teaching Ethics in a Pluralistic Society – A Groundwork for a Common Morality', *Hastings Center Report* **8** (5), 32–39.
5. Daniels, E. J.: 1954, *I Accuse Kinsey*, Christ for the World Publishers, Orlando.
6. Geddes, D. P. (ed.): 1954, *Analysis of the Kinsey Reports on Sexual Behavior in the Human Male and Female*, Dutton, New York.
7. Gurman, A. S., and Klein, M. H.: 1984, 'Marriage and Family: An Unconscious Male Bias in Behavioral Treatment?' in E. A. Blechman (ed.), *Behavioral Modification with Women*, Guildford, New York, pp. 170–189.
8. Halleck, S. L.: 1971, *The Politics of Therapy*, Jason Aaronson, New York.
9. Hare-Mustin: 1978, 'A Feminist Approach to Family Therapy', *Family Process* **17**, 181–194.
10. Holder, A. R.: 1982, 'Case Study: Contraceptive Research: Do Sex Partners Have Rights?', *IRB: A Review of Human Subjects Research* 4 (2), 6–7.
11. Kaplan, H. S.: 1974, *The New Sex Therapy*, Brunner/Mazel, New York.
12. Lennane, K. J. and Lennane, R. J.: 1973, 'Alleged Psychogenic Disorders in Women: A Possible Manifestation of Sexual Prejudice', *New England Journal of Medicine* **288**, 288–92.
13. Margolin, G.: 1982, 'Ethical and Legal Considerations in Marital and Family Therapy', *American Psychologist* **7**, 788–801.
14. Masters, W. H., *et al.*: 1977, *Ethical Issues in Sex Therapy and Research*, Little, Brown, Boston.
15. Masters, *et al.*: 1980, *Ethical Issues in Sex Therapy and Research*, Vol. 2, Little, Brown, Boston.
16. Masters, *et al.*: 1970, *Human Sexual Inadequacy*, Little, Brown, Boston.
17. Rice, D. G. and Rice, J. K.: 1977, 'Non-Sexist "Marital" Therapy', *Journal of Marriage and Family Counseling* **3**, 3–10.

18. Roberts, B. M.: 1981, 'The Use of Surrogate Partners in Sex Therapy', in D. A. Brown and C. Clary (eds.), *Sexuality in America*, Pierian Press, Ann Arbor.
19. Rosenfeld, A.: 1980, 'Inside the Sex Lab', *Science Digest* (Nov/Dec), 102–124.
20. Strupp, H. H.: 1974, 'Some Comments on the Fallacy of Value-free Psychotherapy and the Empty Organism: Comments on a Case Study', *Journal of Abnormal Psychology* **83**, 199–201.

SECTION III

RELIGION, MEDICINE, AND MORAL CONTROVERSY

PAUL D. SIMMONS

THEOLOGICAL APPROACHES TO SEXUALITY: AN OVERVIEW

Theological approaches to sexuality place moral perspectives on sexual conduct, attitudes, and orientations in the overall context of a theological system. A construct or pattern of thought is developed that allows one to relate moral norms regarding sexuality to basic theological affirmations. This overview attempts to provide a succinct analysis of the conceptual framework of various contemporary schools of thought. It makes no pretense of being comprehensive since there are important differences between theologians within similar traditions, and other approaches could be discerned that are worthy of separate treatment. Several basic approaches are here examined in an effort to show why and how theological perspectives are important in sexual ethics.

THEOLOGY AND SEXUAL ETHICS

Sexuality refers to the entire range of issues in the relationships between the sexes. Sexual identity, male-female roles, marriage, celibacy, the purposes of coitus, abortion, contraception, homosexuality, and varieties of sexual play are all involved. The problem for ethics is the question of the norm(s) by which sexual matters are to be judged as to their moral value. By what criteria are sexual issues to be judged right or wrong? Of interest are not only the stated norms or criteria but also the theological base on which they rest.

What is noteworthy is that: (1) theologians may reach similar conclusions about particular issues based on very different lines of reasoning, and (2) strong differences of opinion rest on one or more profoundly different beliefs at theological and philosophical levels. This makes the process of reasoning or the theological-ethical method of great interest and concern. Several base points or areas of thought will be dealt with in any systematic approach. These include theological beliefs,

Earl E. Shelp (ed.), Sexuality and Medicine, Vol. II, 199–217.

philosophical assumptions, style of moral reasoning, and empirical data. Beliefs at the "ground-of-meaning" or theological-philosophical level will largely determine the types of moral judgments one brings to sexual matters. Theology is to ethics what belief is to action. Thought structures provide the conceptual framework out of which human action and moral convictions flow. Thus, it is important to examine the underlying assumptions of various theological approaches.

NATURAL LAW THEOLOGY

Natural law theory has dominated theological approaches to sexuality in the Western Christian tradition. Roman Catholic theologians have most consistently developed this approach, though it can also be found in Anglicanism and Protestantism through the influence of Luther and Calvin.

Natural Law ethics seeks norms that are universal and absolute to give guidance for questions in sexual morality. These principles or norms for conduct are morally binding, since they reflect the truth of God. They are not given by special revelation, however; they are "natural" and thus available to all people. Presumably this establishes a basis for agreement on matters of morals and law among Christians, secularists and people of other world religions.

The roots of this approach lie in Greek thought and can be traced to such thinkers as Heraclitus, Cicero, and Aristotle. The Stoics saw natural law as the demand of the *logos*, which indwelt both people and the cosmos. Aristotle distinguished between the "special law" and the "general law" – those unwritten principles that are supposed to be acknowledged everywhere. He assumed that such "laws" were available through nature, which made them universal. Cicero made it explicit: "There is indeed a true law, right reason, agreeing with nature, diffused among all men, unchanging, everlasting . . . "

Christian thinkers adapted such Hellenistic approaches to science and philosophy to Christian theology and biblical motifs. Augustine's argument that God as pure mind had created people "in his image" as rational creatures shows this influence. His "order of being" ruled out the reality of evil *as such* (it was nothing created), identifying it rather with choices for anything less than the perfect contemplation of God, which was the chief purpose of human existence. The rational powers of the human creature made possible the (rational) knowledge of and communion with God (mind to Mind). Choosing any other action or possibility in the scale of being is to make a choice that is evil. The goal is not evil, for all that is created is good. But choosing less than the highest is an evil choice.

The possibility of human perfection is lost because of the Fall. Because of the original sin, evil now attends all human choice and action in the form of concupiscence, according to Augustine. This is not "lust" in the sense of sexual passion but "desire" or "the compulsion to obtain or possess" that people may direct to any of a multitude of possible goals or objects of choice. This is an irrational component that now attends every act of coitus and constitutes the sin of sexual intercourse. This original sin is passed on to successive generations through conception and birth. Every person inherits Adam's sin because each is a product of coitus.

Since coitus itself is sinful because of the "evil accompaniment", it must have a justifying cause. This, said Augustine, was procreation, which God allowed because of the need for progeny. As long as coitus is open to impregnation, its sin is forgiveable.

This profound negativism characterizes all traditional natural law theory about sexuality. Some of the early Church Fathers sounded more Manichaean than Christian in believing Adam and Eve were sexless in the state of perfection and, had it not been for the Fall, procreation would have been by some harmless mode of vegetation. Augustine argued that Adam and Eve could have willed the sex of their children and would have enjoyed perfect, i.e., rational and controlled, bliss in intercourse. The sex-sin-conception relation was so powerful that the doctrine of the Immaculate Conception had to be developed to further enforce the view that the Virgin Birth was the basis for the sinlessness of Jesus. Logically, of course, the theory would require pushing this argument all the way back to Adam and Eve. Within the early church, celibacy was regarded as the highest calling, while coitus was thought degrading and revolting. Marriage was, at best, a *remedium concupiscentiae*– a remedy for sin – and a sign of sinful lust. That clergy should be celibate was required by the early fifth century. The same theological assumptions continue to undergird the notion of an unmarried priesthood.

Aquinas developed natural law theory most explicitly and systematically. He combined an understanding of the natural state of man with his reading of the purpose of coitus in nature. Following Augustine, he argues that, in the original state of perfection, human beings were perfectly rational, capable of freely willing and thus ordering sexual actions and their consequences. His ordered mind developed a consistent and extensive casuistry – a legalistic code for questions of sexual morality – that formed the basis for pastoral guidance in the confessional. Two basic criteria emerged for judging sexual acts: procreativity and rationality. The two hang together, of course, since "reason" came up with the conclusion that procreation is the primary purpose of coitus. This was confirmed, by this logic, by the observation of the role of coitus in nature, that is, among the animals. Aquinas reasoned that sexual acts could be divided into the categories of "natural" and "unnatural". Natural acts consisted of those that observed proper species and thus were open to procreative possibilities, i.e., human heterosexual coitus. Any sexual act that was not "procreative" was "unnatural". His list of sexual sins was derived by these criteria. Thus, unmarried coitus, contraception, incest, and rape were "natural" sins, while fellatio, cunnilingus, anality, homosexuality, masturbation, and bestiality were "unnatural". Furthermore, natural sins were venial while unnatural vices were mortal, condemning one to damnation. This line of reasoning led to the astonishing notion that masturbation was a more heinous (mortal) sin than was rape.

Problems with this approach have been recognized by leading Catholic scholars to the point that it is now discredited among all but very traditional thinkers. American Roman Catholic scholars have made significant shifts from traditional approaches and now stand at considerable distance from official church dogma. Writers like Charles Curran, Daniel Maquire, and Rosemary Reuther are making creative and positive contributions to contemporary thought. The study by Kosnick provides not only a positive approach to sexuality but a firm ground for significant agreements between Protestant and Roman Catholic thought. The Vatican II document, "The Church in the Modern World", encouraged such creative efforts by acknowledging that the purpose of marriage goes beyond the procreative aspects, and attempted to recover a note of joy for married sex. Even so, official Catholic dogma still rests on Aquinas' casuistry. Pope John Paul II's pronouncements against contraception, a married clergy, abortion, and the ordination of women all reflect the medieval approach to natural law theory. He has declared that, on these questions, the church's teachings would never change.

The fatal weakness of this approach can be seen, first of all, at a very practical level. Natural law theologians disagree and disagree sharply among themselves on issues in sexual morality. While claiming the same basic approach and committed to the same foundational assumptions, conclusions about sexual issues are extremely varied. Roman Catholics are deeply divided on issues from the control of human fertility to abortion – the two primary issues at the heart of the contemporary debate. The point is that gaining agreement on what is universally prescribed and discoverable by reason is practically impossible. Studying nature (the "is") leads to no discernible norm (the "ought") regarding sexuality except variety. Promiscuity and polygamy can as reasonably be argued from nature as can fidelity and monogamy. Nature is a poor guide for sexual morality among people.

Even so, the appeal of basing moral judgments on a "natural law" persists. Every theologian would wish that moral judgments could be based on universal principles discoverable by reason and confirmed by nature and Scripture. Such an authoritative base would "settle the question" presumably and provide confidence and certainty in the midst of shifting social values and relativistic approaches. It would also provide a solid basis for Christians' cooperation with secularists in such matters as social policy and civil law. However, as developed in the Middle Ages, it is also dogmatic and legalistic. Cooperation with secular thinkers ends when opinion divides. The *Syllabus of Errors* made it clear that truth is not to be found outside official church teaching. Since the realm of the supernatural prevails over and is superior to the realm of the natural, and since the church is the keeper of the truth of God, institutional leaders turn to the leverage of law to coerce conformity to official teachings. Such efforts deny diversity and liberty of opinion and thus create antagonism and division among religious groups.

A further problem lies at the theoretical level. Following Aristotle's metaphysics, God is portrayed as pure mind and the *imago dei* is defined as rationality. The notion of original perfection as a state of pure reason does violence to the biblical story and evolutionary understandings of human origins. Human sexuality is better understood on its own terms than on the basis of an abstract construct based on metaphysical speculation. People may be capable of a certain degree of reason, but the evidence is that irrational components also attend human thought and action. This is not due to some primeval "Fall" but to the nature of finitude itself. Natural law criteria are arbitrary norms based on a false understanding of human nature, that is, beliefs about Christian anthropology.

For reasons like these, natural law theory has been largely rejected in theological circles as an adequate approach to human sexuality. The desirability of discerning a "universal" law above "human" laws continues to make it attractive, however. Whether this approach can be satisfactorily re-established remains to be seen. Its associations with repressive, imperialistic, and legalistic code morality may be an insurmountable obstacle.

PROTESTANT FUNDAMENTALISM

Fundamentalists among Protestant theologians reflect and retain many of the theological-ethical perspectives of natural law theory. Theologically, the idea of an original state of perfection is maintained, as is the notion of the biological transmission of original sin. Thus, the Virgin Birth is indispensable for theology as a necessary basis for the sinlessness of Jesus. Even so, infants are not damned; it is personal sin that condemns, though people are not able not to sin. The scholastic stress on rationality is retained in fundamentalism's approach to revelation as propositional truth. The basis of authority is located in Scripture, which is regarded as infallible and inerrant, a repository of propositional truths and explicit moral rules. A doctrinaire, dogmatic approach to religious belief is focused on certain doctrines taken to be the essence of Christian faith and basic to orthodoxy.

Approaches to sexual morality move beyond theological structures to include elements of social customs and what are regarded as nature's laws. The Puritan influence is seen in a strict, legalistic attitude toward non-married coitus. Married coitus is regarded as a good gift from God to be enjoyed, however. Marriage, not procreation, is regarded as the right-making norm for all sexual activity. Procreation is encouraged as is fertility control – a significant departure from traditional Roman Catholic theorists. However, their attitudes and approaches to abortion are virtually identical, since both rely on similar ideas about God's relation to natural processes and notions of soul-infusion that now emphasize the moment of conception as the equivalent of personhood. The strength of tradition over Scriptural authority is apparent at this point.

The complementarity of the sexes is taken as a "natural" norm making all homosexual acts "unnatural" and thus immoral. Biblical passages condemning certain types of homosexual activity are taken as universal and absolute condemnations of any homosexual relation.

Women are also regarded as inferior to men. Eve was responsible for the sin of Adam and bears greater punishment for the Fall. Men are to be in authority and dominant in church and home. Thus the ordination of women is opposed with the same urgency that a hierarchical, husband-father dominated marriage and family are advocated.

The lack of a scientific basis for many of its assertions is among the great weaknesses of Fundamentalist sexual ethics. Its primary problem, however, lies in a failure to develop a theological base for moral perspectives, assuming that certain passages in the Bible explicitly teach what they hold to be true. Stressing a rationalistic, Scholastic approach to religious faith and belief leaves little room for relating the moral life to active obedience. In sexual matters, a stern legalism has been built on the twin towers of human sinfulness and social custom. A legalistic code morality has its counterpart in rationalistic belief in propositional truth.

At a practical level, the possibility of joyful and uninhibited sex within marriage (which is advocated) is greatly diminished by repressive tactics and negative attitudes toward unmarried sex. The high regard for marriage is also undermined to a considerable degree by placing heavy pressures on couples to conform to hierarchical family patterns. This approach tends to emphasize authority for authority's sake rather than encourage the mutual human interactions that make family living truly reciprocal. Family structures should develop out of human and personal dynamics, not preconceived patterns and roles imposed from a romanticized past and rationalized as if they were universally and clearly revealed in God's word.

The strength of this approach lies in its lively sense of the presence and providence of God and in its strong advocacy of fidelity between married partners. Even in this area, however, a rigid authoritarianism tends to enforce fidelity by moral sanctions rather than teach fidelity through strong ties of love, communication, and mutuality. Couples tend to stay married for reasons unrelated to quality of relationship.

"ORDERS OF CREATION" THEOLOGY

The "Orders of Creation" approach to sexuality is associated with theologians such as Karl Barth, Emil Brunner and Helmut Thielicke. These Neo-Reformation or Neo-Orthodox theologians developed ethical thought by taking seriously both the findings of science and the authority of Scripture. They stressed the role of revelation above reason and openly opposed both Scholastic (rationalistic) theology and natural law ethics.

The Reformation emphasis on justification by grace through faith was of central importance for their understanding of the radical nature of faith as obedient response to the living Lord. The believer is directly responsible to God, unable to avoid immediate accountability by obeying (but hiding behind) any number of rules or laws, no matter how valuable such rules might be. The "ground of meaning" belief in the nature of faith as obedient response determines their understanding of the type of obedience required of Christians. Rules or principles may be helpful, but they never adequately summarize the requirement of God directly communicated to the believer.

Another feature of this theological method is the use of paradox, especially in Barth and Brunner. Two contradictory statements could be posed while affirming the truth of each, though not necessarily with equal weight. One could thus say both "yes" and "no" to a certain moral attitude. While some regard this as a type of linguistic chicanery, others perceive it as a way of conveying the complexity of reality. The method has the advantage of avoiding both a harsh, uncritical legalism and a normless relativism. Absolutism is avoided at both extremes of the moral spectrum.

At the level of personal faith, ethics is highly relativistic. Casuistry, or setting down rules for moral rightness (wrongness) beforehand, is excluded as a possibility by the very nature of faith. As Brunner put it, "no universal law can anticipate what [responsible love] means in a world confused and corrupted by sin" ([4], p. 199). Moral seriousness is derived from the fact that one is being addressed by the Eternal God who is Creator, Judge, and Redeemer. The first and primary meaning for Brunner of the notion of "order" is as the direct address or "command" (*das Gebot*) of God. This requires immediate response and thus that the believer be open and prepared to hear God's address in all its particularity. The varieties of action required by God in Scripture can thus be understood. They are to be taken seriously as what God commanded *in situ* of the person of faith. That does not mean that everything in Scripture is normative for the contemporary believer.

What appears to lead to anarchy at the personal level is checked at the social dimension. The nature of persons as social creatures is here joined to the theological notion of human sinfulness. A highly pessimistic theology of human nature (radical sinfulness) requires that constraints be set around personal behavior. This "social principle" was based on a second meaning of God's "order", which was to be found in the "Orders of Creation" (*Schoepfungsordnung*). These are forms or structures given by God in Creation within which the believer lives and which must be accepted as a basic (moral) framework in which to live the life of faith. The "orders" are established by God both to set boundaries for behavior because of the need for the constraint of sin and to provide a means of growth toward the end intended by God. Thus, they are given by God both as Creator and as Redeemer. In the sexual realm, marriage is the "order" that provides the social context in which human sexual behavior is to be expressed.

Brunner believed this "order" is based both in the biblical revelation and in human need. Sexual needs are rooted both in biology and society. People are sexual-social creatures, fashioned by God with capacities for intimate fellowship with others and provided with a matrix (marriage) within which both sexual and personal growth needs can be met. Brunner's ambivalence toward sex can be seen both in his recognition that "without sexuality there can be no full humanity" ([3], p. 365), and his equally strong statement that shame attaches to all sexual activity. He derides celibacy based on the denial of the goodness of sex as non-biblical and non-scientific. Even so, sin makes sexual powers particularly subject to destructive behavior and the exploitation of others. The freedom of grace may be perverted into the liberation of the flesh. Such twistedness has its source in the Fall but now attaches to all sexual congress. This negative element in coitus caused Brunner to reject outright the possibility that Jesus was married.

Two principles thus emerged as norms in Brunner's thought by which all sexual acts are to be tested: love and marriage. God always commands love, but, for sexual activity, only in the context of marriage. For this reason, Orders of Creation approaches are not rightly theologies of sexuality, but theologies of marriage. This structure offers couples the possibility of growth in grace and of personal enrichment. This is the context in which sexuality may be expressed most perfectly. This is God's intention and purpose. Even so, marriage is no guarantee that this redemptive purpose will be realized. Rape, pain, exploitation, and passionate lust may and will attend coitus even within the order of marriage. At least, however, marriage provides a positive constraint.

The evil that attends coitus is not given full reign or license. Even more positively, marriage is a school of the virtues, offering people the possibility of learning the meaning of fidelity in love.

Marriage is thus an instrument of grace. God gives permission for every person to enter the ranks of this order with all its responsibilities and commitments. In marriage one may discover the blessedness intended by God. But striving is still necessary to discover God's riches and his will for fidelity in relationship. Thus, Brunner could affirm the *idea* of marriage as the divine law without making marriage a requirement for every person. Celibacy is acceptable as a calling *above* the calling of marriage. Furthermore, a particular marriage never corresponds exactly to God's intention for marriage. Human institutions reflect the divine Order but also embody human sin. Each marriage participates in the Order given by God, but each falls short of the divine intention.

Roman Catholic theologians rightly saw "points of contact" for dialogue with "Orders of Creation" thinking. While the basis of authority and notions of human nature differ considerably, both rely on a reading of nature and its laws, and their conclusions about sexuality are notably similar: a strong affirmation of marriage, the negative element in coitus, the notion of marriage as a school for the virtues, the virtual indissolubility of marriage and a condemnation of non-marital and non-heterosexual coitus. Brunner does not treat homosexuality explicitly, but the logic of his thought would lead to condemnation. Barth strongly condemns homosexuality as corrupt emotion and perverted physical desire. Thielicke says it is a "borderline situation", treating homosexual orientation as a matter of arrested personal and social growth. Brunner condemns masturbation because it is private gratification, not shared satisfaction with a partner in a context of commitment and fidelity.

The strengths of this approach are considerable. The emphasis on the grace of God and the centrality of love are solidly biblical and prevent the loveless legalism that makes any sexual act unforgiveable and marriage a lifelong bondage allowing of no divorce. There is an affirmation of the goodness of sex, thus rejecting a non-biblical asceticism that developed in the early church. There is also an effort to understand the positive role of marriage in Creation and Redemption without requiring it of all persons.

Problems with this approach are equally apparent. Beginning with a theological construct based on a type of Platonic idealism that posits certain "orders" to be accepted and maintained is artificial at best. Starting with marriage instead of human sexuality is also problematic. The "orders" approach leaves people more concerned to conserve structures than to serve and love people. Their reading of the "laws of nature" was also terribly conservative, reflecting social context and personal history more than a scientific understanding of human nature. Thielicke unapologetically sets forth a double standard of morality: men are polygamous, women monogamous, *by nature*, he says ([16], p. 84). Brunner also accepts inequalities between the sexes in social structures and roles as basic. These are supposedly established by God in creation. Such an approach fails at the same points as natural law theory, *viz.*, the artificiality and unacceptability of the philosophical framework, the lack of a scientific approach to sexuality and the rigidity involved in the roles and relationships of the sexes.

SITUATION ETHICS

While some would question whether "situation ethics" has a theological base, it merits mention because of its impact in the midst of the sexual revolution. The primary exponent is Joseph Fletcher, whose *Situation Ethics: the New Morality* [7] provided a type of apologetic for Christian perspectives in a time when sexual activity was being loosed from all norms. His situationism was an amalgam of existentialism, pragmatism, personalism, utilitarianism, and Christian *agape*.

The one absolute, for Fletcher, is *agape*, for God is love. Thus, in moral matters, the one norm is love, which grants freedom and requires responsibility. He cites the primacy of the love commandment in both Jewish and Christian traditions in support of his view. He reacts strongly against legalism based on rule-oriented absolutes, whether in natural law or Bible-oriented traditions. But he tries to avoid total relativism by positing the rule of love, which was always to be obeyed but could never be reduced to one or any number of secondary rules or codes.

Masturbation, adultery, non-married sex, homosexuality, and various types of sexual play all were acceptable as long as they were loving and thus contributing to human wholeness. By that standard, on the other hand, many types of actions would be wrong – rape, incest, child molestation, and sado-masochistic acts would be contrary to love, for they are injurious to people. He once told a college audience that unmarried sex may be better than married sex. His point (widely misinterpreted) was that marriage was no license to injury or sexual exploitation. Being married did not make rape or psychological manipulation for sexual favors right. Only what is loving is what is right. Thus, an unmarried relationship of love would be better than a married but unloving relationship.

Knowing what love (God) requires is also highly situational, for him. One could not know what is required in advance or standing outside a concrete case. This case-oriented approach (neo-casuistry) he bases on a personalistic ethic (people above principles) and an existentialist epistemology – a type of intuitionism. The wisdom (rules, principles, social mores) of society may help but they must not be used to hinder one's doing the radical act that responsible love may require. One must discover and do what love demands in direct confrontation with the neighbor's need and with all the information at one's disposal. Such is the burden and terror of human moral responsibility.

Fletcher was positivistic in his theological foundations, not bothering to develop systematically the theological groundwork upon which he was working. That does not mean it is without a rationale. He actually developed the first pole of Brunner's paradoxical approach to moral questions. Without ever saying so, he popularized the "personal principle" side of Brunner's understanding of the Divine Command, extending it to include all people, not just Christian believers. He rejected as too wooden and conservative the "Orders of Creation" side of Brunner's thought, which did not suit Fletcher's more optimistic view of human nature.

Critics have rightly pointed to the problem of looseness and relativism in situation ethics. Fletcher's high estimate of human nature made no place for a stress on the radical nature of sin and its effect on all human relationships. However, it is unfair to claim that this is no ethics at all. Fletcher underscored in a way not often appreciated the radical nature of the demand to love. The unexpected, unanticipated call to follow heroic paths in spite of personal risk is at the heart of the love command. This is not easily understood in sexual contexts, however, because of the human tendency to confuse love and lust. In this important area, Fletcher does not give sufficient guidance.

PROCESS THEOLOGY

Process theology has developed Christian perspectives showing the influence of the philosophical thought of Alfred North Whitehead. The central Christian affirmation that God is love is wedded to an understanding of the world that supposedly reflects scientific perspectives of nature and its processes. Evolution is a central metaphor by which all of nature – including human beings – is understood and thus becomes an important hermeneutic for interpreting the Bible.

Whiteheadian thought is characterized by its notion of reality as Being-in-process. It takes account of the subjective aspects of consciousness and the objective reality of matter in such a way that continuity rather than radical difference exists between the two. Energy or spirit permeates everything as Einstein's formula suggests ($E = mc^2$). Consciousness, at least at extremely elementary levels, is present in every living thing. Furthermore, all things cohere together, each affecting and being affected by the other. All of reality is in process. It is not now what it once was and it shall become what it now is not.

Those process theologians who attempt to be orthodox Christians hold that God is both transcendent and immanent. His providence sustains and directs the entire process toward his own future or end. He is the Creator who expresses his very being through creation and, in turn, is affected by the processes of history and nature. God is thus both Being and Becoming, since the creature affects and participates in God's future. Here is a type of panentheism – God is within but also transcends cosmic processes.

The influence of this line of reasoning and its implications can be found in the sexual ethics of thinkers such as Daniel Day Williams, Norman Pittenger, and James Nelson. For them, the biblical revelation that God is love and that people are made for love is the "monarchical principle" in the organization and patterning of human existence. This contention is confirmed in nature – love is the highest response of which any creature is capable. Discovering love makes community possible and is the highest moral achievement.

Human sexuality is understood primarily not from transcendent and eternal principles or patterns but "from below" – in personal life. A phenomenology of the sexual life is the necessary starting point ([18], p. 219). The disciplines of biology, psychology, and anthropology all aid Christians to come to terms with Scriptural insights as to the power and possibilities of sexuality in all its dimensions. Thus, biblical notions of psychosomatic wholeness are close to the working model of what it means to be human among process theologians. The body-soul dualism that afflicts natural law theory is criticized and overcome. The person is a body-soul, an organic and total unity in which rational, affectional, and evaluational aspects are integrated and at the same time grounded and conditioned by physiological processes and material body.

For process thought, sexuality is not an addendum to being human but integral to biological, psychological, and spiritual wholeness. The human ego – the "I" – is distinctively sexual. Maleness or femaleness belongs to personal identity. Sex is to be neither disparaged nor glorified, but cherished and experienced in distinctively human dimensions. Asceticism is neither biblical nor consistent with the needs of human nature. Sex is good in creation and intended for human well-being. Celibacy can be accepted as a way of life in which one fruitful and natural kind of experience is renounced for the sake of service to God and neighbor. One cannot renounce one's sexuality but can sublimate sexual drives.

Sin is the distortion of human sexual identity and the thwarting of God's intention for people in their need for love. This is why decisions about sexual actions are important. They are moral in nature because the person's development may be enhanced or thwarted through sexual experience. The person can be released to love as one is intended to love or can become unloving and thus fall short of God's will and intention.

Love is the moral norm by which all sexual acts are to be measured. Certainly other guides to moral action or general principles can be developed by reflecting on the nature and requirements of Christian love and its relation to human wholeness. Pittenger indicates several features that will identify sexual relations as loving: commitment to the other as a person; fidelity to promises made and agreements entered; mutuality of interests and agreement on acceptable types of sexual play; non-injury to the partner; responsibility for one's actions and for the well-being of the partner; and treating the other always as person, never as a means for the gratification of passion [13]. Control is as necessary as freedom.

Love requires that constraints be set around the types of attitudes and actions that are acceptable for Christian sexual conduct. These are derived both from the human understanding of love and from biblical norms. People have developed unique capacities for love through evolutionary processes. This shows both their kinship to God and their moral obligation to one another. Obeying the requirements of love is both possible and the promise of salvation. The person can be released for love as one is intended to love. This is what redemption, reconciliation, and atonement are all about in the biblical story.

Variety in sexual orientation is accepted as a basic phenomenon of nature. Thus, Pittenger says it is "Time for Consent" toward homosexual relations that are loving and thus moral in nature. Every person, whether heterosexual or homosexual, has the same basic problem – how to discover in relationship one's own identity and experience the power of the love of God for which each is created. All are subject to the same distortions in logic and misdirected drives of passion. Both homosexuality and heterosexuality are manifest throughout nature and are to be accepted as a given of the human condition. The heterosexual majority should not treat the homosexual minority with contempt or scorn in the name of being moral. Such attitudes betray the love for which we are made and deny a fundamental fact of human nature. Sexual orientation is the result of nature-nurture patterning, not of perverse reasoning or a demented morality.

Thus, process sexual ethics does not lapse into a moral relativism. This is not an "anything goes" morality. Control is necessary if sex is to serve the purposes of love and not be a destructive force in the person's life. The enormous power of sex to heal or hurt, bless or destroy, strengthen or weaken the persons involved is to be taken seriously. What one becomes is the result of decisions made along the way. Thus, no one can take a cavalier or casual attitude toward sex. It is too important to be left to the pornographers. Sheer sensuality is a serious distortion and dreadful twisting of the need to discover love and be loving. Even so, sex should not be given ultimate importance. Sexual sins, like those of greed, avarice, gluttony, and hatred, are subject to forgiveness and healing. They should not be singled out as more heinous than other transgressions.

Critics have attacked process thought on both theoretical and practical grounds. Whether Whiteheadian philosophy is amenable to Christian theology has been a major point of contention. The notion of God as personal and transcendent seems compromised by the process notions of God as immanent love and spirit. Certainly those concerned to stress the omnipotence and omniscience of God are disturbed by the "power-as-love" emphasis and the idea of the di-polar nature of God. What such critics betray, of course, is a bias in favor of traditional metaphysics. Whether Whitehead's thought is less Christian than Aristotle's may be subject to debate. At the practical level, critics are unhappy about the lack of absolute rules beyond that of "love" and the general openness toward homosexual relationships in process thought.

The strengths of this approach are considerable. Scientific and religious understandings are integrated rather than polarized or compartmentalized. The insights of each are taken as ways of better understanding the will of God for the world and His creatures. Further, many biblical motifs have been captured, which have been neglected by traditional metaphysics. Without compromising the nature of God as love and goodness, process thought has given fresh insights into the meaning of the suffering and patience of God in creation and redemption. Questions of moral and natural evil are also dealt with as features of an unfinished universe and the struggle of persons to overcome the distortions of which their search for love is capable. Thus, human freedom and responsibility are underscored profoundly in the creature's relationship to the Creator. Scientific world views are taken as aids to understanding the biblical revelation, thus overcoming slavish literalisms based on pre-scientific cosmologies. In sexual ethics, the biblical stress on the goodness of sex as a gift of God to be celebrated is recaptured, as is the notion of the sexual nature of personhood. The will of God is thus understood in terms that are related to human well-being and development, not in terms of abstract or even specific rules that are given arbitrarily and are unrelated to human fulfillment.

SUMMARY AND CONCLUSIONS

Analyzing theological approaches to human sexuality by examining certain base points or underlying assumptions and the moral conclusions to which they lead in this manner causes certain features to stand out. These indicate both their differences and similarities. They can be indicated briefly by way of summary and conclusions.

(1) *Each attempts to discern the importance of "nature" for moral perspectives in sexual matters.* Nature is construed very differently, however, ranging from the interest in a primeval state of nature-in-perfection to an interest in understanding human psychological and

biological needs. The most problematic approach is that which attempts to discover "laws" for moral rules among people from observations of genital activity among non-human subjects. Any adequate or acceptable approach will interpret human sexual needs, capacities, and relationships in all their variety in the light of the biblical revelation. The insights of the sciences will be indispensable in that task.

(2) *All theologians use the Bible and claim its authority to one degree or another.* Protestants are more likely to set biblical authority above tradition, while Roman Catholics will ordinarily interpret Scripture in the light of Church teaching. Fundamentalism's claim about biblical authority in sexual matters is more than a bit suspect, since Puritanism and social mores are more apparent than scholarly approaches to biblical perspectives on sexuality. Neo-Orthodox thinkers take seriously the variety and unity of the Scriptures in a credible manner and have helped restore confidence in Scripture as authority. Process thinkers are less inclined to claim any *source* as authority, believing that authoritative perspectives are and can be only those that are rooted in truth and understanding. For them, God's truth is the truth of reality and all sources of knowledge are helpful in discerning the Will of God. Theologians like D. D. Williams believe their arguments consistent with Scripture, though they make no claim that this is the only possible interpretation or the only possible approach for Christians.

(3) *Each approach uses rules in dealing with sexual morality.* They disagree as to the nature and types of rules that are normative. Traditional natural law theorists held that there were absolute rules grounded in both nature and revelation (whether biblical or special). Aquinas' rationalism developed an extensive system of specific rules by which to measure sexual conduct. These were regarded as binding on the faithful. Fundamentalists are equally oriented toward absolute rules, which they feel are revealed in Scripture. The rules of Neo-Orthodoxy are highly flexible and always measured by love, the only absolutely binding rule. Situationism works with only one rule (love), while admitting others may be helpful. Process thought develops rules that should guide or characterize sexual relationships. These are neither absolutes nor grounded in revelation, however. They are derived by reflecting on the types of actions and attitudes that contribute to human wholeness in sexual relationships.

(4) *Various philosophical perspectives influence the direction and content of theological and ethical reflections.* Natural law theory is linked

to Aristotelian metaphysics; existentialism, personalism, and neo-platonism are reflected in the Orders-of-Creation theologies; situationism shows the influence of existentialism, pragmatism, utilitarianism, and personalism; and the philosophy of A. N. Whitehead provides conceptual models for process theologians. Ties between theology, ethics, and philosophy are both inevitable and necessary, since moral action guides need to be rooted in an understanding of reality that reflects an informed and credible scientific base. Opting for Aristotle seems unacceptably dualistic and tends to continue the battle between science and religion that seems an intellectual cul-de-sac. The great advantage of process thought lies in its monistic world view that not only overcomes the sacred-secular, body-soul dualism of traditional thought, but offers a conceptual framework consistent with Hebrew (biblical) perspectives.

(5) *Each approach will continue to have committed advocates for reasons rooted in tradition, authority, belief systems, and personal experience.* The perception of truth is not a purely rational enterprise, but one that involves a variety of subjective factors that do not lend themselves to simple isolation, analysis, or modification. People believe as they do for various reasons, one of the most powerful of which is simply the tradition in which they learn basic approaches to theological questions. What is required is for each to respect the others and continue to be open to unfamiliar perspectives. Good faith toward others will be necessary to avoid the religious imperialism that attempts to impose its perspectives on those of differing approaches and convictions. All are engaged in the quest for the truth of God, which requires at a minimum the humble recognition that all human understandings are limited. In the important area of sexuality all people are attempting to discover not so much fidelity to rules as fulfilment of the need to love and thus embody the power and presence of God. Surely the role of the church is to facilitate the need to discover the uniquely human capacities for love, fidelity, and forgiveness in community. Insofar as it contributes to the person's ability to relate sexual powers to God's redemptive purposes, it becomes a creative and indispensable agent of the grace of God.

Southern Baptist Theological Seminary,
Louisville, Kentucky, U.S.A.

BIBLIOGRAPHY

1. Abbott, W. M. and Gallagher, J. (eds): 1966, *Documents of Vatican II: The Church in the Modern World*, G. Chapman, London.
2. Barth, K.: 1937, *Church Dogmatics*, Vols. II.2 and III.4, T. and T. Clark, Edinburgh.
3. Brunner, E.: 1937, *The Divine Imperative*, O. Wyon (trans.), Westminster Press, Philadelphia.
4. Brunner, E.: 1970, *Love and Marriage*, Collins, London.
5. Curran, C.: 1982, *Moral Theology: A Continuing Journey*, University of Notre Dame Press, Notre Dame.
6. Curran, C.: 1978, *Issues in Sexual and Medical Ethics*, University of Notre Dame Press, Notre Dame.
7. Fletcher, J.: 1963, *Situation Ethics: The New Morality*, Westminster, Philadelphia.
8. Kosnick, A., *et al.*: 1977, *Human Sexuality: New Directions in American Catholic Thought*, Paulist Press, New York.
9. Long, E. L., Jr.: 1967, *A Survey of Christian Ethics*, Oxford University Press, London.
10. Macquarrie, J.: 1970, *Three Issues in Ethics*, SCM Press, London.
11. Nelson, J.: 1978, *Embodiment*, Augsburg, Grand Rapids.
12. Pittenger, N. W.: 1968, *Process Thought and Christian Faith*, James Nisbet and Co., Ltd., London.
13. Pittenger, N. W.: 1972, *Making Sexuality Human*, Pilgrim Press, Valley Forge.
14. Pittenger, N. W.: 1973, *Love and Control in Sexuality*, Pilgrim Press, Valley Forge.
15. Pittenger, N. W.: 1970, *Time For Consent: A Christian's Approach to Homosexuality*, SCM Press, London.
16. Thielicke, H.: 1964, *The Ethics of Sex*, John W. Doberstein (trans.), James Clarke and Co., Ltd., London.
17. Thielicke, H.: 1966, *Theological Ethics*, Vol. I, William H. Lazareth (ed.), Adam and Charles Black, London.
18. Williams, D. D.: 1968, *The Spirit and the Forms of Love*, James Nisbet and Co., Ltd., London.

JAMES J. McCARTNEY

CONTEMPORARY CONTROVERSIES IN SEXUAL ETHICS: A CASE STUDY IN POST-VATICAN II MORAL THEOLOGY

More than twenty years now have passed since the Bishops of the Roman Catholic Church first assembled in Rome to participate in the Second Ecumenical Council held at St. Peter's in the Vatican. Since that time, the religious practices, if not the beliefs, of millions of Catholic Christians worldwide have been significantly modified and changed. One only has to reflect on the widespread abandonment and ignorance of the Church's heretofore hieretic and sacred language (Latin) by contemporary clergy to grasp the profundity of the change that has taken place. The Bishops of Vatican Council II called for a reexamination of every aspect of the Church's life and teaching so that its message of salvation could be more clearly understood by contemporary men and women.

In the Vatican II "Decree on Priestly Training" (*Optatam totius*), we find the following mandate: "Other theological disciplines should also be renewed by livelier contact with the mystery of Christ and the history of salvation. *Special attention needs to be given to the development of moral theology*. Its scientific exposition should be more thoroughly nourished by scriptural teaching . . . " ([10], p. 69).

This Conciliar command for the renewal of moral theology has been taken very seriously by contemporary theologians both in the United States and abroad. This theological *aggiornamento* has been documented very thoroughly over the years by Richard A. McCormick, S.J., in his "Notes on Moral Theology" section of the journal *Theological Studies*, several years of which have now been drawn together into a book [10].

Updating, however, has not been without its difficulties. As in the case with almost all seriously undertaken human endeavors, a certain

Earl E. Shelp (ed.), Sexuality and Medicine, Vol. II, 219–232.

dialectical movement is discernible in contemporary Roman Catholic moral theology, which has spawned many controversies and disagreements in its wake. In this article I would like to describe the basic starting points of this dialectical interaction and then use contemporary controversies in Roman Catholic sexual ethics as case studies in order to describe the dialectical nature of this interaction and to discover the future direction of Roman Catholic ethical thinking on a variety of topics.

THE CONFLICT

Most disputes in contemporary ethical theory center around deontological or duty-based ethics versus teleological or outcome-based ethics. In duty-based ethics, it is claimed, one arrives at a certain set of right-making characteristics of human actions either from reason, divine authority, or some other source, and applies these to human actions. Certain goods or benefits flow from these actions and activities, but it is the right actions themselves that are ontologically prior to the goods or benefits, such that even when goods are not immediately forthcoming one is still obliged to do one's duty and do what is right.

Teleological, or outcome-based, ethics looks first at the goods or benefits that are derived from human action and tries to discover what right-making characteristics bring these about. In this system, the production of goods and benefits, or the avoidance of evil outcomes, is ontologically prior to determining the rightness or wrongness of a particular action or a class of actions.

Much has been written about the conflict between the right and the good, or between deontological and teleological ethics, and I do not intend in this brief article to review that literature. Suffice it to say that there is a strong deontological element in contemporary Roman Catholic theology; nevertheless, its traditional tools have always been grounded in some sort of teleological approach or another. The conflict I perceive in Roman Catholic moral theology is not the conflict of teleology versus deontology, but two conflicting approaches that are both essentially teleological. These two approaches have been described as "prudential personalism" on the one hand and "proportionalism" on the other ([1], pp. 167–169).

PROPORTIONALISM

Proportionalism is an approach to Roman Catholic moral theology that accepts the traditional thesis in Catholic thought that the morality of a human act is determined by its object, its circumstances, and the intention of the agent. However, proponents of proportionalism believe that it is not possible to look at the objective content of the act alone without considering at least the intention of the agent, in order to determine its morality or immorality. Proportionalists hold that looking at the action apart from the agent only determines its ontic, or pre-moral, goodness or badness, not its morality. Moral good or evil can only be posited when there is an acting agent. "For proportionalists, the object of a human act, apart from its circumstances and the agent's intention, is only a premoral object which involves certain human values and disvalues. It becomes a moral object, which is morally good or bad, only when consideration is given to the proportion of values to disvalues involved in the concrete act taken in the circumstances, and in view of the intention of the agent" ([1], p. 161).

Thus the proportionalists would deny that there can be any actions that are intrinsically evil; that is, actions that by their very nature are immoral, irrespective of the intention of the agent. Nonetheless, several of these scholars would agree that there could be some actions, for example, the brutal torture of children, that would be so personally evil that no proportionate value could ever be adduced for their performance. Proportionalists are quite comfortable with agreeing that it is the end and only the end, properly understood, that justifies the means.

Proportionalists would also reject the traditional principle of double effect as formulated in Roman Catholic moral theology. They would hold that the first condition of the double effect principle, namely, that the action in itself be either good or indifferent, can only be determined by an analysis of the proportionality of the good effects versus the evil effects in the performance of the act as intended by the human. Thus they would claim that the principle of double effect can be reduced to the principle of proportionality ([14], pp. 170–173).

For this approach, clarification of values in the subject and adherence to those values in conflict situations is more important than rigid adherence to an objective law. Proportionalists believe that the structures of human nature give us hints as to the rightness and wrongness of certain human actions and may be used as *prima facie* guides for morality, but that they are not absolutely binding under every circumstance. What one must look at, for proportionalists, are the relative values and disvalues for oneself and for others stemming from the performance of a given act or a class of acts before determination of their rightness or wrongness can be made.

Several Roman Catholic moral theologians in America belong to this methodological school, including Richard McCormick, S.J., of Notre Dame University, Charles Curran of The Catholic University of America, and Timothy O'Connell of St. Mary of the Lake Seminary, Mundelein, Illinois.

Nonetheless, it has been pointed out that it is very difficult to reconcile this theory with established Christian teaching, which continues to be followed by recent Popes and the magisterium of the Church, that there are some kinds of acts, e.g., direct abortion, which are always morally evil no matter what the circumstances or the intention of the agent ([1], p. 162).

PRUDENTIAL PERSONALISM

Prudential personalism is a term that has been used to describe the teleological approach of those who, while emphasizing that Christian ethics can never be reduced to a simple matter of duty, nonetheless see the development of the human person in its fullest relation with others and with God as the criterion and object of human morality. For prudential ethics, morality ultimately is not a matter of obeying rules, but of intelligently seeking appropriate, concrete behavior by which to achieve human personal and communal goals ([1], pp. 162–164). Prudential personalists want to relate morality to the issue of how human actions in concrete contexts contribute to the growth of persons and community. In raising this issue, they want to include not only the individual person in his or her self-becoming, but also include all the finalities and purposes they envision as gifts from God that make us human, and are not ultimately able to be chosen or refused on our own authority.

Prudential personalists teach that people need to understand the God-given nature of human reality and also the specialness of each human person. This understanding cannot be achieved through reliance on a single principle alone, such as love, justice, pleasure, or any other value, but by the development of many principles reflecting the complexity of human personal life. From these principles, certain values are developed that help us formulate moral rules by which we are to make prudent choices. Through these moral rules, people inform their consciences concerning particular moral choices in a prudent manner. "Such a moral logic is, therefore, 'prudential' in its practical intelligent effort to reach goals, and it is 'personalist' in that it works not for superficial goals but for the total realization of the inherent needs of the human person in community" ([1], p. 167).

Prudential personalists hold that acts premorally considered, that is, antecedent to the intention of the agent, already possess a natural orientation and finality which a free human act, if reasonable, must expect. Thus they disagree that moral theology can be reduced to a simple analysis of proportion of value over disvalue in given acts or classes of acts. Prudential personalists do not disregard the importance of proportionality in determining the morality of some actions, but they would hold that there are some actions that can never be performed, no matter what the proportion of values over disvalues might be determined to be.

Theologians in this school, such as Rev. Benedict Ashley, O.P., Rev. Kevin O'Rourke, O.P., Germaine Grisez, and William May of The Catholic University of America, all contend that they are much more within the mainstream of Catholic thought and teaching in the area of ethics, and especially in sexual ethics, than those who espouse proportionalism. We will now investigate the applications of these two theories to the realm of sexual ethics to see the different positions that adherence to one of these two schools can bring.

HUMAN SEXUALITY

In a book written several years ago, Eugene Kennedy points out that compassionate religious leaders could do more to integrate human sexuality in people's lives than an army of sexual scientists, but that they have to give up their instincts to control men and women in order to achieve this [7]. I agree with this insight and believe that the most positive contribution that Catholic moral theology will make to human sexuality is not in a discussion of the controversies, which I shall present momentarily, but with a new development now taking place, which looks not so much at sexual actions to determine their rightness or wrongness, but looks at human sexuality in the context of intimacy, spirituality, and interpersonal growth. I believe that the experience of the Christian community has much to say on these issues and that the energies of moral theologians in the Church would be much better spent in discussing the importance of the relation of love and physical embodiment, human love as a symbol of divine love, the meaning of friendship and commitment, and the improvement of the quality of family life, rather than focusing on whether premarital sex, homosexuality, and masturbation are always and everywhere wrong. There are hopeful signs that this is taking place [13], but unfortunately much ink and passion is still spilled and spent over the morality of concrete sexual actions, without taking the context of human sexuality and spirituality into proper consideration.

MASTURBATION

Perhaps the clearest contrast between the approach of proportionalism on the one hand and the prudential personalism on the other can be illustrated through the human action of masturbation. Proportionalists, by and large, are able to find many justificatory reasons for self-induced sexual stimulation leading to orgasm, among which are obtaining semen for fertility testing or for diagnosing certain venereal infections, obtaining reasonable relief from excessive sexual tension, or even preserving fidelity in marriage when physical separation is a necessity. Proportionalists like Charles Curran even maintain that the amount of disproportion or negative value in masturbation is such that it would constitute a very minor infraction of the moral law at worst (in Catholic moral theology this would be referred to as venially sinful).

Not that proportionalists would be willing to justify any and all masturbatory behavior; there are times and situations in which masturbation is wrong. "Masturbation simply for the sake of the pleasure involved, without any effort at control or integration, can be indicative of self-centeredness, isolation, and evasion of relational responsibility. . . . Exploitation of one's sexuality, freely, deliberately, consistently in this manner creates a serious obstacle to personal growth and integration and constitutes substantial inversion of the sexual order – an inversion that is at the heart of the malice of masturbation" ([8], pp. 227–228). Proportionalists are more inclined to look at the context of masturbatory action rather than at the act itself. Most proportionalists would not want to see masturbatory behavior used as a general substitute for the natural sexual urge and would find this a disproportionate value, were it to come to pass; nevertheless, they would be open to many positive values that can be brought about by autoeroticism. As Norman Pittenger states: "Masturbatory practices are wrong only in context; they may be right, even desirable, when all possible relationships with others are cut off and the only opportunity for relief of strong irresistible sexual drives is through self-stimulation. Cases differ, as do situations" ([16], p. 76).

Prudential personalists, on the other hand, are more inclined to see masturbation as an intrinsically and serious disordered act because "the deliberate use of the sexual faculty outside normal conjugal relations essentially contradicts the finality of the faculty. For it lacks the sexual relationship called for by the moral order, namely the relationship which realizes the full sense of mutual self-giving and human procreation in the context of true love" ([17], pp. 9–10). As Monsignor Miguel Benzo of the Pontifical University of Salamanca states: "The serious moral disorder of masturbation is based fundamentally on its character as an imperfect and unsatisfactory sexual act, since it excludes the essential orientation to serve as the language of love and the means of procreation" ([17], p. 109).

Because these prudential personalists see the natural and divinely constituted orientation of the human sexual faculty towards both procreation of the species and towards an expression of interpersonal love, this purpose cannot be changed by human fiat. Thus, they would hold that each and every act of autoeroticism is, by its nature, immoral. They claim to discover, rather than create, the natural end of sexual activity as a given in human life, ordained by nature and sanctioned by God with certain ends or purposes that cannot be changed. These ends and purposes are for the human good and are thus teleological, but they are not contingent on human values and preferences. Thus, these goals and goods are revealed by nature itself and not invented or created by human willing. This approach is much simpler to present than that of the proportionalists, because if an action is, by its very nature, against that which is divinely constituted to be for our ultimate good, then it is immediately disproportionate in the objective order and no other discussion has to be presented in terms of subsidiary outcomes. Actually, many proportionalist theologians view the prudential personalist approach as a simplistic understanding of what it means to be a human being in the eyes of God.

HOMOSEXUALITY

Proportionalists deal with the phenomenon of homosexuality in a variety of ways. Some proportionalists, like John McNeill, see homosexuality as a given, and attempt to discover purposes and values that the reality of homosexuality brings with it. Far from seeing homosexuality as always somehow disvalued, McNeill and others who write in a similar vein believe that homosexuality has positive value both for the individual homosexual and also for the community at large. Some of these values include greater sensitivity to the opposite sex, more concern for the poor and destitute of society, and a greater appreciation by homosexuals for the spiritual and religious dimensions of human life.

Greater sensitivity to the opposite sex is achieved precisely because there is little sexual attraction to the opposite sex, and relationships proceed, not in terms of sexual intimacy, but on the basis of mutual respect, friendship, and collaboration. Concern for the poor and destitute arises from the fact that gay people are generally part of a persecuted and minority group and feel bonds of solidarity with other such groups in society. Finally, it was Jung who believed that the androgynous character of homosexuality, in his terms the "ascendency of the feminine", allows for a greater spiritual sensitivity and religious meaning for the homosexual person, especially for the gay male. For proportionalists such as McNeill, these positive values are more than sufficient to offset the more negative such as rejection from society, the inability to have children, and the inability to have an opposite sex spouse and normal family constellation. These proportionalists would see homosexuality as comparable to lefthandedness, i.e., a minority within the population but in no way better or worse than the heterosexual population at large.

Other proportionalists, such as Charles Curran, pursue the matter of homosexuality differently. Curran teaches that the ideal meaning of human sexual relationships is in terms of male and female. He sees opposite sex complementarity as the goal towards which human relationality should move. But he admits that, because of the presence of sin in the world, ideal structures are not always accomplished in practice and that there are some who are legitimately homosexual in an exclusive way, i.e., constitutive homosexuals. Curran maintains that, while ideally it would be better to enter into heterosexual relationships that have the possibility of life-giving love as part of the sexual structure, nevertheless, it is permissible for constitutive homosexuals to enter into exclusive long-term relationships that are sexually expressive. He justifies this by his theory of compromise, which implies that we must take the world as we find it and derive what value we can from broken existential situations, even though this value is less than perfect. Thus, Curran sees a positive value in committed love between two persons of the same sex when it is impossible for them, for one reason or another, to express or experience sexual love with a person of the opposite sex. Curran's reasoning for this is that it is better to enjoy some of the fruits of human sexuality than none at all, and he believes it is the same reasoning that allows the Church to permit a heterosexual couple, one of whom is known to be sterile, to marry. Curran's approach on homosexuality could be described as making the best of a bad situation ([4], p. 118).

Prudential personalists also vary widely on their approach to homosexuality, even though all would agree with the statement that "according to the objective moral order, homosexual relations are acts that lack an essential and indispensable finality. . . . Homosexual acts are intrinsically disordered and can in no case be approved" ([17], p. 9). After having accepted this basic premise, there are theologians such as John Harvey who, while agreeing that homosexual orientation is morally neutral, nonetheless would encourage self-control through asceticism and a celibate existence as the only possible response, in the Christian context, to homosexual orientation. Harvey would emphasize the re-education of the homosexual person in the meaning of love and intimacy so that positive real relationships can be developed without the need of genital intercourse ([3], pp. 148–150).

A slightly different emphasis is found in the *Introduction to the Pastoral Care of Homosexual People* [2], a document written by the Catholic Bishops of England and Wales, which, while accepting the principles set forth in the Vatican Declaration of Sexual Ethics, nonetheless highlights many of the values of same sex orientation excluding genital intimacy. The authors of this pamphlet encourage friendship, closeness, intimacy, and affection between homosexuals, knowing full well that this might occasionally bridge over into genital expression. Their approach is that, while this genital expression is wrong, many subjective elements may enter, which lessen culpability, possibly, every time it happens.

More important even than the morality of homosexual acts is the morality of attitudes towards homosexuals. This has been dealt with explicitly in a pastoral letter written by the Bishop of Brooklyn, Francis Mugavero, in his letter entitled "Sexuality – God's Gift". He states the following: "Our community must explore ways to secure the legitimate rights of our citizens, regardless of sexual orientation, while being sensitive to the understanding and hopes of all involved" [13].

Thus, while prudential personalists believe that we must maintain the natural orientation of the sexual faculty at all costs, many are becoming increasingly sensitive to the fact that persecution of others has much more disvalue associated with it than occasional misuse of the sexual function. It is a hopeful sign that this emphasis is coloring much of the writing that is being done about homosexuality today by proportionalists and prudential personalists alike.

PREMARITAL SEX

Proportionalists tend to make a number of distinctions when discussing premarital sex. In considering sexual activity between those engaged to marry, proportionalists generally take a very open view. Francis Manning suggests that in these cases perhaps marriage has already taken place in terms of the mutual consent, even though it has not been formalized through liturgical ritual or official ecclesial witnesses. "Like most of life's decisions, becoming married is not an instantaneous action, but a process that takes time. At a certain time in the process coitus becomes an appropriate expression of the love that exists and the will to place all that one is in the service of the other. . . . As a general rule of thumb, however, it might be suggested that the couple should have manifested to others their sincere intention to marry, and that the ceremony itself is not too far distant" ([8], pp. 160–161).

Norman Pittenger distinguishes between casual sex and committed sex in the following way:

Everybody is familiar with the fact that young people are likely to wish to engage in what is called sexual experimentation. There is nothing evil about the desire, as such. The problem is essentially one of respect and genuine personal concern. The young man out to "make" every girl he meets shows scant respect for those victims of his desires, nor does he have much if any concern for their personal integrity. There can be occasions when a boy and girl genuinely feel a deep affection for each other and wish to cement that affection by physical relationships. Condemnation is easy here, especially by older persons in whom the physical urge is less strong. What is required is an awareness of the strength of the pressure felt by one or both of the parties, with the recognition that there can be cases where love, or what is honestly thought to be love, almost demands the physical act. No rules can be laid down, especially in a day when "petting" in various degrees of intimacy, is taken for granted by our culture. But it can be said that no young boy should take advantage of another human being, unless and until the other is ready and willing, while no young girl should let herself be easily put in the position where she cannot refuse consent precisely because she has "led on" her friend, or has allowed him to proceed, stage by stage, to the point where there is no holding back ([16], pp. 54–55).

The authors of *Human Sexuality* suggest that premarital sexual morality is largely a matter of drawing honest and appropriate lines concerning premarital sexuality. They ask:

Is it self-liberating, other-enriching, honest, faithful, life-serving, and joyous? To what extent, if any, is there selfishness, dishonesty, disrespect, promiscuity, the danger of scandal or of hurting or shaming family and loved ones? Young people in particular should be challenged to question themselves as to whether they are simply not using each other to prove their respective masculinity, femininity, sexual attractiveness or prowess. To this should be added for young and old alike the question of willingness to accept a child if one is conceived ([8], p. 168).

Thus we can see that proportionalists, by and large, see that there can be some positive values to premarital sexuality, including that of getting to know the intended spouse better, expressing one's intimacy and affection for a special other in one's life, and even showing a dimension of intimacy and caring that cannot be expressed in any other way. They would, however, caution against premarital sexuality that is exploitative or casual, and would caution that this practice, when engaged in frequently, tends to depersonalize and objectivize human relationships to the point where they become merely functional.

Proportionalists generally tend to approach premarital sexuality in a processive and developmental way, seeing that premarital sex, when engaged in by committed couples, is less problematic than the so-called "one night stand". Most proportionalists tend to take a rather sympathetic stand towards premarital sexuality, yet one that emphasizes the life-giving and other-enhancing qualities of human intimacy expressed in a sexual way.

Prudential personalists, on the other hand, tend to take a much more conservative and rigorous attitude toward premarital sexual expression.

Their position can be summed up in the following quote from the *Declaration on Sexual Ethics*:

> Today there are many who vindicate the right to sexual union before marriage, at least in those cases where a firm intention to marry and an affection which is already in some way conjugal in the psychology of the subjects require this completion, which they judge to be connatural. This is especially the case when the celebration of marriage is impeded by circumstances or when this intimate relationship seems necessary in order for love to be preserved.
>
> This opinion is contrary to Christian doctrine, which states that every genital act must be within the framework of marriage. However firm the intentions of those who practice such premarital sexual relations may be, the fact remains that these relations cannot insure sincerity and fidelity, the interpersonal relationship between a man and woman, nor can they protect this relationship from whims and caprices. . . . Sexual union therefore is only legitimate if a community of life has been established between the man and woman. . . . The consent given by people who wish to be united in marriage must therefore be manifested externally and in a manner which makes it valid in the eyes of society ([17], p. 7).

O'Neill and Donovan, while adopting the prudential personalist approach that sexual intimacy is limited to the married state, nonetheless come a long way toward presenting a very developmental view of human sexuality and suggest that intimacy must come in stages. They believe that engaged couples must get to know one another in terms of physical intimacy as well as verbal and psychological intimacy and suggest that going too far in terms of intimacy is better than not going far enough. They do advise, however, that when certain intimate actions are known to lead to genital expression, they should be avoided in the future.

All in all, the prudential personalists feel that the genital expression of sexuality is limited by nature and divine mandate to the formally married state, and that any expression of sexuality outside this state does not have the full life-giving and love-sharing significance that it ought to have.

OTHER CONFLICTS

At this point, I could present other examples such as contraception, abortion, divorce and remarriage, adultery, and sexual perversions as other areas in which the proportionalists and the prudential personalist views would differ. However, I believe that by now the basic methodology of these two schools within Catholic moral theology has been distinguished clearly enough. Generally, the prudential personalist view is more conservative, more oriented towards a natural law approach, more traditional, and generally the one formulated by the official teaching authority of the Church, including Popes and Bishops. The proportionalist approach, on the other hand, looks for more subtle distinctions, deals on an existential rather than a theoretical plane, and is more likely to allow for exceptions. I believe that the most urgent task awaiting Catholic theology in the next decade is the bridging of these two approaches in some creative way that allows for the recognition of the tradition while at the same time seeing the benefits of the proportionalist analysis. Until this is done, much confusion will remain in Catholic minds, since it is generally the proportionalists who publish much more in moral theology, while ordinary parish priests, pastors, and bishops adopt the prudential personalist approach. My conviction is that, appearances to the contrary, the two approaches are not irreconcilable, but that a great deal of work needs to be done in order to bring them within a common understanding of the Christian moral life. It is hoped that out of the creative dialectic of these two antithetical approaches (both teleological, I would contend), a new creative synthesis will emerge that includes the Christian wisdom of the past yet deals realistically with the problems of intimacy experienced by people today.

St. Francis Hospital,
Miami Beach Florida, U.S.A.

BIBLIOGRAPHY

1. Ashley, B. and O'Rourke, K.: 1982, *Health Care Ethics: A Theological Analysis*, 2nd ed., The Catholic Health Association, St. Louis.
2. Catholic Social Welfare Commission, Catholic Bishops of England and Wales: 1981, *An Introduction to the Pastoral Care of Homosexual People*, New Ways Ministry, Mt. Rainier.
3. Curran, C.: 1983, 'Moral Theology and Homosexuality', in J. Gramick (ed.), *Homosexuality and the Catholic Church*, The Thomas More Press, Chicago, pp. 138–168.
4. Curran, C.: 1972, 'Sexuality and Sin: A Current Appraisal', in M. Taylor (ed.), *Sex: Thoughts for Contemporary Christians*, Image Books, Garden City, pp. 104–121
5. Curran, C. and McCormick, R. (eds.): 1979, *Readings in Moral Theology No. 1: Moral Norms and Catholic Tradition*, Paulist Press, New York.

6. Dominian, J.: 1971, *The Church and the Sexual Revolution*, Dimension Books, Denville.
7. Kennedy, E.: 1972, *The New Sexuality: Myths, Fables and Hang-Ups*, Image Books, Garden City.
8. Kosnik, A., *et al.*: 1977, *Human Sexuality: New Directions in American Catholic Thought*, Paulist Press, New York.
9. May, W.: 1976, *Sex, Love, and Procreation*, Franciscan Herald Press, Chicago.
10. McCormick, R.: 1981, *Notes on Moral Theology: 1965 through 1980*, University Press of America, Washington.
11. McNeill, J.: 1983, 'Homosexuality, Lesbianism, and the Future: The Creative Role of the Gay Community in Building a More Humane Society', in R. Nugent (ed.), *A Challenge to Love: Gay and Lesbian Catholics in the Church*, Crossroad, New York, pp. 52–64.
12. Milhaven, J.: 1970, *Toward a New Catholic Morality*, Image Books, Garden City.
13. Mugavero, F.: 1976, *Sexuality – God's Gift*, The Chancery, Brooklyn.
14. O'Connell, T.: 1976, *Principles for a Catholic Morality*, The Seabury Press, New York.
15. O'Neil, R. and Donovan, M.: 1968, *Sexuality and Moral Responsibility*, Corpus Books, Washington.
16. Pittenger, W. N.: 1970, *Making Sexuality Human*, Pilgrim Press, Philadelphia.
17. Sacred Congregation for the Doctrine of the Faith: 1977, *Declaration on Sexual Ethics*, United States Catholic Conference, Washington.

ROBERT H. SPRINGER, S.J.

TRANSSEXUAL SURGERY: SOME REFLECTIONS ON THE MORAL ISSUES INVOLVED[1]

THE SEXUAL ISSUE

The medical achievement of successful transsexual surgery makes it imperative that we reconsider the relation between ethical theory and moral practice. Those individuals are considered transsexual who are anatomically of one sex but feel themselves to be of the opposite gender. Such individuals do not doubt their anatomical identity. They are convinced it is the wrong one and, consequently, desire their anatomy to correspond with their felt gender identity. Their condition can become acutely painful, leading to severe depression, disruption of their lives, and even suicidal tendencies. By transsexual surgery we are referring to surgery undertaken physically to reconstruct a person biologically of one sex but claiming to be psychologically of the other. The techniques are available to construct cosmetically satisfactory, and often functional, genitalia.

The opinion of moral theologians concerning transsexual surgery has evolved significantly over the past thirty years. The earliest opinion claimed that such surgery was not morally permissible because it would constitute "a grave mutilation of the human body." This opinion was based on three premises, all of which have been subsequently challenged. The first is that gender identity is determined by the genitalia. The second is that the ethical norm involved in a decision concerning surgery should be exclusively " . . . the body's health and integrity" ([7], pp. 133–135). The third premise is that any surgery not required for the good and integrity of the body violates humanity's custodianship of the body and represents an immoral act of supreme ownership. This opinion is challenged in the first part of this essay.

At a later period of time Ashley and O'Rourke have no objection to surgery performed on true hermaphrodites, i.e., on persons whose biological sex is ambiguous, where the purpose is to improve sexual

Earl E. Shelp (ed.), Sexuality and Medicine, Vol. II, 233–247.

identity in favor of what seems to be the predominant sex of the person [1]. However, they continue morally to condemn transsexual surgery where there is no biological ambiguity. Expanding the principle of totality and integrity to include not only the good of the whole body but the good of the "whole person", these moralists can conceive of the possibility that transsexual surgery could be for the overall good of the person and, consequently, not violate the principle of totality and integrity. However, they continue to reject this type of surgery on moral grounds as " . . . intrinsically outside the limits of ethical medicine." The reason they give for this evaluation is that such surgery has as its purpose " . . . not genuine treatment of a psychological illness but an illusory adjustment involving a destructive loss of bodily integrity." This moral judgment rests on the presupposition that gender dysphoria is a psychological "illusion" that can be cured by psychotherapeutic means without recourse to surgery ([1], pp. 318–319).

A more recent evaluation of transsexual surgery is to be found in *The Handbook on Critical Sexual Issues* [9]. The authors of this book concede that: "If it were possible for a person objectively to be of one gender while at the same time of the opposite anatomical and physiological sex, then in such a situation hormonal and surgical alterations could be seen as corrective rather than being an unjustifiable type of mutilation in the theological sense." However, the authors reach the somewhat surprising conclusion that ". . . at the present time there seems to be no convincing evidence that such a condition can exist" ([9], p. 170). Precisely what the nature of the "objective" evidence is and how it could be proven, the authors do not specify. The insistence on the need of such objective evidence seems suspect, since we are dealing with cases of gender dysphoria with a psychological phenomenon, which of its very nature is based on subjective experience.

The authors themselves display a curious ambivalence in their conclusions. At one point they concede: "Both the transsexual and the therapist may well be convinced that the surgery would be genuinely therapeutic and ethically acceptable." And again they state: "Some pastoral counselors . . . may well make such a judgment (i.e., that surgery is morally justified)." However, their final word is: "At this time, no clear and definitive moral decision in favor of transsexual surgery can be made because of lack of convincing proof that such disharmony does objectively exist and that transsexual surgery could be justified as corrective and therapeutic" ([9], pp. 171–172).

What is of primary interest in this history is the evolution of theological moral judgment from total condemnation to an ambivalent acceptance. This evolution has obviously been the product of an accumulation of empirical evidence as to the nature and extent of the problem of gender dysphoria and the failure of traditional means to provide a solution [15].

What Is the Problem?

In order to form an appropriate moral judgment on transsexual surgery, we must be able to answer several questions concerning human psychosexual development and psychological formation of gender identity. How does this identity come about? Is a felt conflict between gender identity and biological givens to be understood as a "neurotic illusion," or can that conflict have a non-illusional factual basis? Finally, and most important, to what extent is psychological gender identity, once fully established, open to change by psychotherapeutic means? Or are there cases where the only hope, or at least the best hope, for harmony between psyche and soma is to be found in transsexual surgery? Finally, what have been the empirically verified results of past attempts at surgical intervention?

Our problem arises from the fact that human psychosexual development, precisely as human, is not subject to purely biological determinism, whether genetic or hormonal in nature. Rather, that development is a long process of many stages and each stage is open to a variety of variations. The overall result of this process is that humans are infinitely various in their sexuality and every person has a uniqueness about him- or herself that needs to be respected [8].

Some experts refer to at least nine stages in psychosexual development. The first five stages are predominantly biological, yet still open to variations. In the first six to eight weeks of development, the embryo is neither male nor female; that is to say, the gonads are neither ovaries nor testicles. A number of genetic and hormonal factors will determine their development.

First stage: Chromosomal sex (xx, xy).

Second stage: Gonadal sex (ovaries, testicles).

Third stage: Hormonal sex (body hair, breasts, other glands).

Fourth stage: Internal sex organs (uterus, prostate).

Fifth stage: External sex organs (clitoris, penis).

The sixth through ninth stages are believed by most experts to be primarily questions of psychological development, although these psychological developments are heavily influenced by the previous genetic and hormonal developments ([12], pp. 476–478).

Sixth stage: Sex of assignment and rearing (what parents want).

Seventh stage: Gender identity (one's inner sense of being male or female).

Eighth stage: Gender role (one's outer enactment of the inner sense of self as masculine or feminine).

Ninth stage: Sexual object choice or orientation (one's erotic attraction to male or female).

Although all the earlier stages obviously play some role, one's gender identity seems to be principally determined by sex assignment and rearing. Experts point out that parents can and sometimes do prefer, even unconsciously, that their child were not of the biological sex with which it was born. These experts claim that the child can clue into the parent's wish, even if unconscious, and adopt the sexual identity that corresponds to that wish. Note that gender identity precedes developmentally gender role and, consequently, is not determined by it. Once the child has a sense of his or her identity as male or female, that child can appropriate whatever range of masculine or feminine characteristics it wishes. This is very much the result of socialization. Sexual object choice or sexual orientation is also developmentally consequent to gender identity. This leads to a complex understanding of various forms, psychological and biological, of homo- and heterosexuality. For example, one may be biologically male and have a psychological gender identity as female. Consequently, if one's sexual object choice is male, one is biologically homosexual but psychologically heterosexual. If the sexual object choice were female, one would be biologically heterosexual and psychologically lesbian.

Conclusions

There can be no legitimate question that there is a substantial population that suffers from gender identity dysphoria to the point where they are willing to consider transsexual surgery. In fact, during the last twenty years there have been over six thousand reported cases of transsexual surgery.

As McCarthy and Bayer recommend, one should never consider

transsexual surgery until all other treatment possibilities have been explored [9]. Extended evaluation with a one to two year trial period prior to the formal consideration of surgery is accepted practice at most reputable centers. To proceed prematurely to surgery without such evaluation would be unconscionable.

First of all, the therapist involved should explore possible explanations of gender dysphoria that would not presume a genuine split between the person's sexual anatomy and gender experience. One must eliminate the possibility that one is dealing with pseudo-transsexualism based on a physical or psychological trauma, e.g., temporal lobe epilepsy, homophobia homosexuality, borderline personality syndrome, etc. Once the physical or psychological trauma is healed, the dysphoria will disappear. To proceed prematurely to surgery in such cases would be disastrous for the individual involved.

Second, more conservative means of treatment, such as psychotherapy, should be tried. However, once gender identity is fully established in an adult, it would seem that therapeutic efforts to change it can frequently be counterproductive, resulting in greater dysphoria, guilt feelings over failure to change, depression, etc. In most cases in my experience, the most productive therapy is that designed to help the person accept and live with his psychological gender identity in the healthiest and most productive way possible. As a result of therapy, many transsexuals will make the choice to undergo transsexual surgery after all other forms of adjustment have been tried and found wanting.

The third stage involves careful screening to determine the depth of the person's gender preference. A trial period of living out the chosen sex role should be initiated. At the end of a one or two year experience of living the new sex role, the person under treatment and the team of specialists may be ready to consider transsexual surgical procedures. Where someone has undergone such a process of evaluation and treatment, it is my opinion that there is little ground for serious doubt that transsexual surgery would be a legitimate moral choice.

The most important question, I believe, that remains to be answered is the empirical question of the effectiveness of this surgery in relieving the suffering of gender identity and leading to a happy and harmonious life. Unfortunately, there is a dearth of empirical studies here to rely on. Of the over six thousand sex reassignment case histories, we have follow-up studies in only a few hundred cases. Meyer and Reter gave us a competent overview of this empirical data in their 1977 article "Sex

Reassignment: Follow Up" [11]. For example, in The Hopkins Study in 1971, there was an attempt to follow up on thirty-four gender identity clinic patients who underwent operations over against a control group of sixty-six unoperated patients. None of the operated patients voiced regrets at reassignment. Here and in every other follow-up reported, by far, the vast majority who underwent surgery reported fair to excellent results in terms of subjective relief. However, there was no objective advantage in terms of social rehabilitation ([11], p. 1014).

In the light of this data, extreme caution is called for before proceeding to surgery. The authors of the "Follow Up" article conclude: "It seems clearly beneficial for patients to be considered for surgery within the environs of an organized group. Abandonment of a program and precipitous requests for surgery are contra-indications for it" (*ibid.*).

Since there is a strong body of professional opinion that gender identity dysphoria is frequently not capable of being cured through psychotherapy, and since empirical evidence supports the judgment that surgical intervention under certain conditions can successfully relieve this dysphoria, I conclude that there is adequate evidence for a probable moral judgment in favor of such surgery.

THE BIBLICAL DATA, THE PARAMETERS OF HUMAN INTERVENTION

Part I began with a reference to "humanity's *custodianship* of the body" and a proscription of "supreme ownership." It is time we examine the rationale of this biblical assumption underlying sexual reassignment. Is this heretofore prevailing assumption valid, namely, that the human is custodian only of soma and psyche? Is there a middle ground in the Judeo-Christian tradition between custodianship and supreme ownership?

> Male and female he created them. God blessed them and said, "Be fruitful and increase, fill the earth and subdue it, rule over the fish in the sea, the birds of heaven and every living thing that moves upon the earth" (Genesis, 1:27–28, *The New English Bible*).

What possible relevance can this text from the second millenium BCE have for the second millenium after? It has much to say, as voices raised from the scientific community testify. A biophysicist addresses the question thusly:

Increasingly the advances being made in many areas of science and technology pose ethical and moral dilemnas which *cannot* be resolved by facts alone. Rather, the proper utilization of our new scientific findings requires that we face up to some terribly critical decisions, based upon our most fundamental values and beliefs ([2], p. 3).

For those of the Judeo-Christian tradition the Bible as a source of values needs no apology. For thinking persons of whatever tradition the splitting of the atom and the discovery of the genetic code have created an openness to an ethic of any source that transcends the technological.

The biblical data here set forth offer guidelines, set limits, and give direction not just to psycho-sexual wholeness but to a broad spectrum of psychological and medical interventions currently in use, e.g., genetic surgery. Indeed, they shed light on the whole technological question of our times ([3]; [4], pp. 215–244; [13]). With such broad scope Part II cannot promise a final go-ahead to all applicable procedures. Its modest claim is to give a perspective and chart an overall course for the human venture into soma, psyche, and protoplasm.

With these preliminaries in mind we take up the book of the sacred writings of Jews and Christians. As one might expect, exhaustive search reveals no passage of specific pertinence to gender fusion or prenatal sex determination. The Bible is not a scientific book, as Copernicus and Galileo have shown. It does reflect the primitive science of its day, but its questions about human well-being, life, and happiness are not our questions. One can, however, tease out of the Bible a religious anthropology and an ethic underlying the problems of our spiritual ancestors and ourselves.

There is the injunction to choose life: "I, the Lord, have put life and death before you. . . . Choose life that you and your descendents may live" (Deuteronomy 30:19, *The Complete Bible*). Have we here a basic principle: what promotes human life and happiness is good and approved by the Creator? This would be to extrapolate the text from its context, characteristic of too much theological argument of the past. "Life" here means survival and the fullness of life in the promised land. The words are spoken to a people in exile and looking toward restoration to their homeland. They say nothing directly of the individual's happiness through restoration of psychic wholeness or of prenatal intervention.

We are faithful to scriptural hermeneutics, however, in extracting from the text an injunction to heed the Torah, the teachings of the Lord.

One chooses life by "loving the Lord your God and by heeding his commands." But here a precaution is necessary. The Bible is not a unified book. The Jewish Scriptures incorporate a variety of traditions, Yahwist, Eloist, Priestly, and Deuteronomic, the four not harmonized but set down side by side. Thus we have two creation accounts in Genesis, each with a distinct message. The New Testament has four Gospels, each with a distinct message. They sometimes contradict one another. An editor of today would reject the manuscript of the Bible as a poorly arranged compilation, prelogical, with conflicting themes. Hermeneutics warns us to beware of fundamentalism, a too literal use of words ([10], pp. 38–42).

Despite this complexity, however, the Scriptures do manifest a religious unity. There are basic themes hidden within the diversity of the various traditions. One overarching theme, germane to our search, is the will of the Lord. Israel is guided in every step of its history by Yahweh's bidding. By the will of God the prophets judged them. By it Moses led them forth from Egypt. Jesus in the New Testament is depicted as passionately devoted to "the will of my Father". No plea by his family, rebuke by his disciples, or threat of his foes could deter him from following it, faithful Jew that he was.

To violate the will of God is the sin of infidelity, hateful to Jew and Christian alike. It is to set the creature in a position of independence from the Creator. This is the ultimate hubris, traditionally called the Fall or original sin. The first question the Bible enjoins on us, whether in releasing the power of the atom or manipulating the genetic code, is: Have we divine authorization?

On this basis many believers have settled the issue of bio- and psychological engineering of today. The following is a sample in point:

> Humanity falsely believes that its life and the life of creation are at its disposal to do with as selfishly desired. The first act of this rebellion is directed toward a tree in the garden. . . .
>
> The threats presented by both nuclear weapons and genetic engineering spring from the conviction that humanity has the right to manipulate, create and destroy creation as it sees fit. Life is its own to possess. By contrast the biblical picture presents the creation as God's gift to be preserved, nurtured and treasured. . . . Such actions *claim prerogatives for humanity that rightly belong only to the Creator* ([6], p. 22).

The argument represents not only Evangelical Protestantism. It speaks for a cross-section of Christianity and, I presume, for Judaism as well.

This position is a respectable one, as the instance of nuclear weapons makes clear. But genetic engineering is another matter. Even here it may prove to be correct. No one can claim to know the mind of God infallibly. These pages offer, however, another reading of the biblical data that differs sharply from the foregoing.

But first let us resume the theme of Yahweh's will to savor its full import. The will bound the Israelites to follow unconditionally in the way set forth for them. Under their great spiritual leaders they bound themselves absolutely, both individually and collectively, to the observance of the Covenant relationship. Its claims on them tolerated no contrary allegiance, whether from the gods of other nations or from human authority, even from their own rulers. Neither the kings of Babylon nor of Israel itself made contrary demands upon them but that the prophets rose up to call them back to obedience to Yahweh's will.

This point needs emphasis. Technological research and development must proceed with caution, honesty, and seriousness of purpose, lest we fly in the face of this underlying biblical mandate. The gods of research subsidy, of medical renown, or of pragmatic necessity must not dictate our decisions.

We come to a crucial point of our inquiry. Does the will specify in the concrete the *content* of human decision *vis-à-vis* human life? Does it specify the applications of the teachings "Choose life" and "Do not kill"?

There are specific provisions as to animal husbandry and agronomy. Leviticus proscribes the crossbreeding of cattle and of plants as a violation of divine order (19:19). To apply this teaching, however, to gene splicing is a farfetched analogy. Reflection on human experience and the advances of science have long since convinced Jew and Christian that hybrid corn and the Missouri mule are not violations of the underlying truth behind this Mosaic ruling. Note that what was seen in one age as contrary to divinely established order is not such in another.

Crossbreeding rules aside then, let us look at the content of divine teaching as a whole and the manner of its imparting. What of the vast body of regulations respecting sexual mores and health in Leviticus and Deuteronomy? Are these foreordained, inscribed on stone tablets from the beginning? The figure of Moses trudging down the mountain with the Tablets of the Law is but a delightful metaphor of the compiler. In reality the Law of the Lord unfolds in the course of their history. It is something to be discovered. Moreover, it is changing, elastic.

Eichrodt, the great commentator on Jewish Scriptures, makes this point. No foreknowledge of God's will, he declares, gives them mastery over an historical decision or provides the right application of the eternal norm to the concrete situation: "Rather, it is into the midst of history with all its insecurity and unforeknowable possibilities that God's will leads. . . . Right decision can be made only if there is constant readiness to listen to the demand which is being made in the situation" ([5], pp. 25–26).

In the light of this hermeneutic principle we may conclude that, even if there be concrete teaching relevant to our subject, its application would be questionable. Scripture does not predetermine technological decision today. Rather, it throws us back on our own resources to be guided only by the spirit of the Law, as were our spiritual forebears.

We of the Western tradition, schooled as we are in the exact mode of thinking of the Greek philosophers and of the Roman lawyers, are uncomfortable with such apparent normlessness. The Torah tells us, "Love the Lord your God with your whole heart and your neighbor as yourself." But when we inquire what this means in the concrete the answer comes back, "Discover for yourselves. Read the signs of the times in the light of Yahweh's love for you and your love for one another."

Our question, then, resolves itself to this: how free are God's people to search for the meaning of the will, to take the initiative, even to make honest mistakes? The ultimate answer to this question is not ethical but ontological. It lies in discovering what man and woman *are* in relation to God and creation, a question prior to the ethical ought. Our relationship to the Lord is the basis of the ethical yes or no as to intervening into human life and our ambience.

Traditionally, the response has been that the human is creature and servant of the Most High. In the New Testament human status is often set forth as stewardship over a householder's possessions. We are administrators of the master's goods, called to strict account and with limited discretionary powers over the oil and wine of the owner, God. The themes of creature, servant, and steward are well known. They have been extensively developed by theologians and taught by the churches and synagogues. They do not, however, by any means exhaust the full stature of the human *vis-à-vis* creation and the Creator.

They represent only one side of the divine teaching, human subordination to God. But there is another side, a godlike dignity of the human. It is expressed in Psalm 8:4–6:

When I see the heavens, the work of thy fingers, the moon and the stars which thou hast formed; what is man that thou shouldst think of him, and the son of man that thou shouldst care for him? Yet thou hast made him but little lower than God, and dost crown him with glory and honor. Thou hast put all things under his feet (*The Complete Bible*).

Nor is this just hyperbole from a lyrical psalmist. It pervades other books as well:

The other side . . . becomes plain in the fact that all the Old Testament witnesses we have mentioned also speak of a unique connection of the incomprehensible and mysterious Creator with his creature. . . . It consists of God honoring man and him alone of all creatures, by addressing him and confronting him as "thou." As One who offers himself for communion he emerges from his hiddenness into a morally positive relation with his creature, in which he shows him his special place and task in the world. . . . As the one whom God's word meets, he comes to God's side and confronts the rest of creation ([5], p. 30).

One source for this biblical teaching is found in Genesis 2:19, "So the Lord God molded out of the ground all the wild beasts and all the birds of the air and brought them to the man to see what *he* would call them; whatever the man should call each living creature, that was to be its name" (*The Complete Bible*).

To get the full sense of this passage one needs to read the whole chapter, undistracted by the enigma of the rib and the symbolism of the serpent. To this beginning one must add the development of this theme in the prophetic books, notably Isaiah, Hosea, and Jeremiah. More than a mere echo of Genesis, these sources exalt the status of the human to that of a people loved by a father, surrounded by a mother's love, even enjoying the tenderness of a husband for his bride. These symbols of Yahweh's relation to the human person contain enormous implications for the power to intervene into human life. To ignore this added dignity and power is to denigrate the creature and the Creator.

Adequate development of this theme would take us through the whole of the Scriptures. For this we have not space. We limit ourselves to the creation theme in Genesis. Chapter 2, cited above, is supplemented in the Genesis 1 account which sets forth:

the effects of man's independent spiritual nature, by which he is set at God's side. In man's destiny as being made in the image of God, the priestly thinker gives utterance to the thought that man cannot be submerged in nature. . . .

The Creator's greatest gift to man, that of the personal I, necessarily places him, in analogy with God's being, at a distance from nature ([5], p. 30).

This exaltation of man and woman translates in theological terms into lordship under God over creation. Another expression of this, according to a longstanding theology, has held that we are co-creators. This truth has been commonly accepted by the churches in the instance of natural generation of human life. It is patent from the power of the gonads with which we are endowed and from the experience of parenting. In other words, we have read this empowerment from human history. Why have we been slow to apply co-creation to the works of our hands and minds, scientific invention, whether in the order of nature at large or of human life?

Male and female the Lord made us and told us, "Increase and multiply." There is no mention here as to the quantity and quality of life. These were left to human reflection and wisdom to set. The same text enjoins that we "dominate the earth." The limits to this are likewise to be set by human standards as to the proper use of environment and resources. The context does make clear that respect for life and all creation should be shown for what God has made and declared to be good. But no greater restriction than this is set by the text to the human task of increasing the quality of life by medical means today.

True, we have not exercised wise lordship over the goods of the earth, as ecology makes clear. In fact we have been egregiously destructive. Nor have we had due regard for the norms of justice in the economic systems we have elaborated, whether capitalism or socialism. We have been untrustworthy junior partners in the promotion of life, as the figures on abortion and the history of warfare demonstrate. But do these gruesome pages of history mean that we must close the door to further psychic or genetic manipulation of life? No, Scripture enjoins rather radical repentance and a new beginning. This response the sacred text expresses through all its pages as the obedience the will requires.

We have been grossly irresponsible in that we have failed to heed the teaching to respect the animal world and the land with the respect that "I the Lord" have shown for my creation. This message is expressed in detail in Exodus 23:19, Leviticus 19:19, Deuteronomy 22:9, culminating in Job 31:38–40 ([5], p. 32).

The Book of Job poses a special problem. It states that the wisdom to hold sway over the universe is God's alone, a thesis contrary to the one here sustained. (See, e.g., Job 28:1–28.) Were we to take the word of Job as final we should have to side with the position cited above, "Humanity neither knows enough about the intricacies of life, nor possesses the moral discernment and spiritual wisdom to design 'more perfect' human beings" ([6], p. 22). In Job's view human reason has become bankrupt of wisdom in the face of evil.

Later sages than Job, however, describe the giving of the Torah, God's revelation, as communicating God's own wisdom to humans. The Book of Sirach is an example. "The Lord himself created her [wisdom] . . . and poured her out upon all that he made; upon all humankind has he chosen to bestow her; he supplied her liberally to those who love him" (1:9–10; cf. also 51:20).

From this too brief survey we may conclude: Scripture proclaims the lordship under God of woman and man over the rest of creation. As for dominion over human life, this is implicit in the creative powers with which God has endowed the noblest of all he made. These powers are not restricted to human generation but include the creations of our hands and minds. The status of the human race is that of bride-lovers of the Lord.

These exalted realities are the ontological basis of the ethical responsibility we bear toward human life and its environment. As yet this dignity has not deeply penetrated the consciousness of the adherents of the Judeo-Christian tradition. We live in an era when servanthood and stewardship still dominate our religious thinking. Stewards and servants have far less power and attendant responsibility than junior partners.

Until lordship-lover consciousness comes to the fore, we are not ready to exercise the increased responsibility it entails. We are not prepared to acquire the wisdom and to exercise the mastery over the whole of creation that Scripture allows and the present state of technology demands. We ought, then, to join ourselves to the prophets in our midst who urge caution and a slower pace of technological development. Until such time, however, biblical teaching does not put a padlock on the laboratory door or demand a recall of the medical technological instrumentalities in use on the grounds that these claim rights over life that belong only to the Creator.

How far we may go in this direction is not *a priori* set down for us. The biblical injunctions remain open-ended. We may indeed proceed farther than stewardship warrants. What "farther" means in detail is to be determined by human wisdom in the light of the great biblical truths we have seen. This much we may say in detail. Sex reassignment, observing the precautions stated above, seems in accord with the biblical teaching of partnership with and under God.

On the other hand, this teaching and human wisdom do not condone amniocentesis for the determination of sex, preliminary to elimination of a fetus of "unwanted" sex. Cost-benefit analysis may not be the sole arbiter for withholding life-sustaining procedures ([14], p. 44). The developed nations may not produce a race of olympic specimens while the rest of the world goes without the nutriments and vaccines necessary for decent human living. This violates the teaching, "Love your neighbor as yourself."

As for future discoveries, love for the neighbor, informed by all the wisdom we command, should shape our decisions. Is a love ethic a vain utopian goal? Many say so in a world sharply cleft into East and West, North and South, where only the message of fear is heard by warlords and self-interest is the primary rule of the ruling powers.

The Bible says we may and ought to strive for a world ruled by love, the heart of the Reign of God promised not just hereafter but in this life.

Crossroads of Reflection and Action,
New York, New York, U.S.A.

NOTE

[1] I am indebted to John J. McNeill, S. J., for many of the ideas expressed in this essay.

BIBLIOGRAPHY

1. Ashley, B. M. and O'Rourke, K. D.: 1978, *Health Care Ethics*, The Catholic Hospital Association, St. Louis.
2. Augenstein, L.: 1969, *Come, Let us Play God*, Harper and Row, New York.
3. Callahan, D.: 1972, 'New Beginnings in Life', in M. Hamilton (ed.), *The New Genetics and the Future of Man*, Eerdmans, Grand Rapids, pp. 90–106.
4. Callahan, D.: 1973, *The Tyranny of Survival*, MacMillan, New York.
5. Eichrodt, W.: 1951, *Man in the Old Testament*, SCM Press, London.
6. Grandberg-Michelson, W.: 1983, 'The Authorship of Life', *Sojourners* 12, 18–22.
7. Healey, E. F.: 1956, *Medical Ethics*, Loyola University Press, Chicago.
8. Lothstein, L.: 1984, 'Theories of Transsexualism', in E. Shelp (ed.), *Sexuality and Medicine: Conceptual Roots*, D. Reidel, Boston, pp. 53–71.
9. McCarthy, D. J. and Bayer, E. J.: 1983, *Handbook on Critical Sexual Issues*, Pope John Center Press, St. Louis.
10. McNeill, J.: 1976, *The Church and the Homosexual*, Sheed Andrews and McMeel, Kansas City.
11. Meyer, J. K. and Reter, D. J.: 1979, 'Sex Reassignment: Follow Up', *Archives of General Psychiatry* **36**, 1010–1015.

12. Money, J.: 1961, 'Hermaphroditism', in A. Ellis and A. Abarbanel (eds.), *The Encyclopedia of Sexual Behavior*, Vol. 1, Hawthorne Books, New York, pp. 472–484.
13. Moraczewski, A.: 1983, 'Genetic Manipulation, Some Ethical and Theological Aspects', in A. Moraczewski (ed.), *Genetic Medicine and Engineering*, The Catholic Health Association of the United States, St. Louis, pp. 101–119.
14. Shinn, R., *et al.*: 1980, *Human Life and the New Genetics*, National Council of Churches of Christ in the U.S.A., New York.
15. Toulmin, S.: 1981, 'Marriage, Morality and Sex Change Surgery: Four Traditions in Case Ethics', *The Hastings Center Report* **11** (8).

SUPPLEMENTARY READINGS

Peterson, M: 1981, 'Psychological Aspects of Human Sexual Behaviors', in Pope John Center (ed.), *Human Sexuality and Personhood*, Pope John Center Press, St. Louis.

Stoller, R.: 1965, 'Passing and the Continuum of Gender Identity', in J. Marmor (ed.), *Sexual Inversion*, Basic Books, New York.

RONALD M. GREEN

THE IRRELEVANCE OF THEOLOGY FOR SEXUAL ETHICS

An often referred-to passage in Nietzsche's *The Gay Science* depicts a madman who descends into a village crying, "God is dead. God remains dead. And we have killed him." To the madman's dismay, however, the villagers stare mutely at him. Finally, he comprehends their silence. "I have come too early," he says,

> my time is not yet. This tremendous event is still on its way, still wandering; it has not yet reached the ears of men. Lightning and thunder require time; the light of the stars requires time; deeds, though done, still require time to be seen and heard. This deed is still more distant from them than the most distant stars – *and yet they have done it themselves* ([24], pp. 181f).

One can question the accuracy of Nietzsche's implicit prophecy about religion in the modern era. In many ways, religious belief is as vital as ever, with even conservative, fundamentalistic forms of religion making a comeback. At the same time, forces at work when Nietzsche wrote, especially the findings of modern science and the value structure of industrial civilization, have eroded religious commitment and belief for many people. But whether Nietzsche was right or wrong about religion in general, there is no doubt that specific elements of traditional religious faith in the West have been thrown open to question by modern culture.

In the remarks that follow I want to suggest that Nietzsche's prophecy is especially pertinent where sexuality is concerned. In my view, *most aspects of the traditional sexual ethics and theologies of sexuality associated with the Western religious traditions may be dead*. They have been killed by technological, social, and cultural changes in our era. It is true that many individuals continue to respect and obey the sexual norms of the Jewish and Christian traditions. It is also true that many religious thinkers and theologians, even if they disagree with specific aspects of

Earl E. Shelp (ed.), Sexuality and Medicine, Vol. II, 249–270.

Western religious teachings on sexuality, continue to believe that the "core" teachings of these traditions merit our continued respect. Nevertheless, it is my contention that all of the elements of these traditions, including some of the common foundational beliefs rooted in biblical faith, may be obsolete.

I make these claims with some hesitation and certainly not in the spirit of a writer of an earlier generation like Bertrand Russell for whom Christian teachings on sexuality were simply preposterous [27]. Many earlier critics who took such a position were dealing with ethical teachings and theologies shaped by centuries of anti-body dualism or by a more recent Victorian prudery, both of which are alien to key insights of biblical faith. In the decades since these critics wrote, advances in scriptural studies and biblical theology have enabled a generation of thinkers to rediscover and represent the far more "naturalistic" and sexually positive dimensions of Hebrew and early Christian faith. Indeed, as a teacher of Jewish and Christian ethics, I have spent a good deal of my career leading students through this newer literature and dispelling some of the misunderstandings of Jewish and Christian sexual teachings that still flourish in popular culture. Nevertheless, despite this important process of repristination and rehabilitation, I believe the biblical tradition's understanding of human sexuality may have become radically out of date. One indication of this is the fact that in order to retain any relevance for this tradition, contemporary theologians have been more and more driven to accommodate it to often contrary information or insights drawn from non-religious sources. While this endeavor may seem praiseworthy to those who believe the tradition must remain relevant (and I agree that such openness to change is far better than a stubborn retention of outdated ideas), the net result is that the religious traditions have ceased to mold culture or even to be informative in this area. They have become, instead, passive recipients of formation by secular and non-religious sources.

The roots of the problem I am identifying can be traced to some of the central ideas about human sexuality associated with biblical faith and with the Jewish and Christian traditions arising from it. These ideas represent fully understandable, even morally commendable accommodations to the social and cultural circumstances that spawned biblical faith and that remained in force for centuries. But since these circumstances have changed so radically in our own day, it is questionable whether these centuries-old ideas retain validity. In particular, I want to single out and examine four separate ideas lying at the heart of biblical thinking about sexuality which have been challenged by developments in our own time. They are (1) the idea that sexual conduct, especially genital sexuality, is per se morally and religiously significant, (2) the idea that human sexuality is properly expressed only in a context of an enduring, lifelong personal relationship, (3) the idea that human sexuality is normatively heterosexual, and (4) the idea that human beings can be categorized and social roles assigned on the basis of gender.

(1) *Sexual conduct and genital sexuality especially as per se morally and religiously significant.* There is no question that biblical faith, and the religious traditions to which it gives rise, have consistently regarded sexuality as an area of consummate moral and religious importance. From the seventh commandment to the prophets' castigations of "harlotry" and fornication, sexual misconduct is regarded as among the gravest moral and religious offenses. It is true that in the New Testament, especially in the well-known episode of Jesus's treatment of a woman taken in adultery (John 8:1–11), there is a suggestion that other forms of sin, especially sins of pride and arrogance, may be more serious than specific sexual transgressions, but there remains a condemnation of sexual immorality itself. Jesus tells the adulteress to depart and "sin no more," and the Apostle Paul, surely representing the spirit of early Christianity on this matter, classifies forms of sexual unchastity and license as among the sins that exclude one from the Christian community (e.g., I Corinthians 5). We are, of course, so accustomed to this understanding of morality that it seems odd to ask why sexual behavior should merit such condemnation. In our culture, shaped by biblical faith, "immorality" and "sexual immorality" have often been synonymous. Yet if we explore the bases of this biblical perspective, it becomes clear that they reside, in part, in technological, social, and cultural conditions that no longer exist. Principal among these is the linkage between sexuality and conception that prevailed up to the dawn of our era. In a context where every sexual act might lead to the creation of a new human life, it is not surprising that sexuality should be subject to strict control and supervision. Sociological and psychological realities work together here to support a stern regulation of sexual conduct. Like many traditional societies, ancient Israelite culture had its major institutional basis in the family. The socialization and protection of the young, employment and material support, possession and transmission of pro-

ductive property and even physical protection from aggression were largely the responsibility of the extended family unit under the control of a male head or patriarch. It is not surprising, therefore, that in this context efforts should be made to discourage conduct that might threaten family stability, family loyalty, or the certainty of bloodlines [26]. The result is the familiar double-standard protecting the rights of married men in their women, but placing lesser constraints on male sexual conduct that does not jeopardize other males' rights. Thus, the capital offense of adultery is defined in the Hebrew bible only in terms of extra-marital sexual conduct involving a married woman (Leviticus 20:10). A married man's liaison with an unmarried woman or a prostitute does not constitute adultery (Numbers 5:12–31; [7]).

Psychological realities conspire to reinforce this standard. In a context where every sexual act is potentially conceptive, sexuality is charged with meaning and forms of power relationship. For a male to sleep with a woman is literally to take possession of her life. What is a moment's pleasure and dalliance for him may alter the course of her existence by imposing forms of deep material or emotional dependency. In this context, sexuality is almost always the assertion of the male's power over the female, and a humane society, limited by technological constraints, will expectedly seek to protect its women here, even if doing so imposes new burdens and limitations on them. No more bizarre illustration of these forces can be found than the biblical law requiring a rapist to pay his victim's bride price and to marry her (Deuteronomy 22:28–29).

I need not detail any further the ways in which the linkage between sexuality and conception in a traditional social context eventuates in the kinds of sexual norms and attitudes with which we are all familiar. The important question, however, is what happens when this linkage ceases to exist or when the social context of nearly total dependency (especially female dependency) on the family is altered? In a different connection, moral philosophers sometimes ask what the effect on our traditional norms regarding violence might be if human beings possessed hard carapaces and were totally immune to physical threat. Would we still believe it is very wrong for persons to assault one another physically? Or, again, if human beings were not mortal, if dying were but a moment's sleep from which we awoke refreshed, would we still regard homicide as a consummate moral and legal wrong? One does not have to be a moral relativist to conclude that profound technical change can alter specific moral conclusions. Although there are basic moral rules that forbid our injuring other persons and that require us to respect others' autonomy and freedom, what constitutes injury or the deprivation of liberty in specific circumstances can be altered by social or technological change.

Precisely this kind of change has come about in the sexual realm. Contraception and the back-up of abortion have profoundly altered the meaning of sexual conduct between consenting adults. The corresponding emancipation of women from material dependency on men and on the family has also changed the character of sexual acts. No longer is every sexual engagement charged with psychological and social significance for its participants. Whether for good or bad, sexual relations have been rendered as potentially casual as any friendly encounter between individuals. At the conclusion of her epoch-making book, *The Second Sex*, Simone de Beauvoir asks whether the traditional understanding of sexual encounter as an act whereby the "passive" woman is dominated or controlled is a necessary aspect of human sexuality. She believes not, and she speculates that in a context where females have essential power over their own lives, a reverse understanding is equally possible. Why should the female's role in intercourse be understood, not as a passive submission to the male, but as an active *utilization*, even *exploitation* of him ([5], p. 809)? Whether de Beauvoir is right or not, it is clear that her speculations presume a context in which the involuntary link between conception and sexuality no longer exists. And it is just that link that has been broken in our time.

I should voice several cautions here. I am not saying, of course, that the availability of contraception and abortion means that sexuality need no longer be of moral concern to us. Many individuals, especially younger people, are unable or unwilling to prevent conception and their often casual exercise of sexuality constitutes one of the gravest moral problems facing our society. (Whether this problem is best handled by re-imposing older norms or attitudes is a question I will not handle here, although I am convinced that in matters of popular attitudes and ethics one can never simply go back). Nor am I saying that sexuality necessarily is or should be a casual matter. Many persons (perhaps most of us) continue to hold attitudes toward sexuality that were shaped by the older cultural norms, and for these people any sexual encounter can be charged with deep personal and moral significance. Then, too, it is possible that sexual experience involves other more abiding interpersonal considerations than those created by its long association with conception. In a moment, we will examine the claims of those who believe that every sexual act creates forms of interpersonal relationship between the sexual partners, which may greatly affect them for better or worse. My point is not that all sexual conduct has necessarily ceased to be a source of moral concern nor that it is morally insignificant, but that a major foundation for that concern and sense of significance, the necessary linkage between sexuality and conception, has vanished. This means that validity of all the traditional religious attitudes resting on this basis are open to question. In our present situation, therefore, it is no less reasonable to hold the view expressed by James R. Smith and Lynn G. Smith that we should today switch from considering sexuality primarily in ethical terms to construing it in an appropriate esthetic context. As the Smiths observe:

> This is not to avoid dealing with normative requirements and ramifications of sexuality but to recognize that with the presence of contraception, affluence, leisure, mobility, a relatively educated middle class, and a host of minor influences on sexual decision-making, the kind of consequences which follow from sexual interaction are more akin to a living art (which ideally engages all of the sensibilities as well as the senses) than a moral problem ([30]), pp. 38f.).

(2) *That human sexuality is properly expressed only within the context of an enduring, lifelong personal relationship*. Some would contend that it is unfair to characterize the biblical perspective on sexuality as being founded solely on a concern with the procreative and social consequences of sexuality. While this concern is undoubtedly present, there is also a deeply *personal* view of the nature and implications of sexuality. In the Bible, it is argued, sexual intercourse is regarded as an intense form of communion between persons. It involves a special form of self-revelation and mutual "knowing". Because of this, sexual acts are never merely casual. They have enduring consequences for the partners, and it is for this reason that the Bible condemns sexual promiscuity and insists on the location of sexual expression within the context of the enduring, lifelong personal relationship of marriage.

This understanding of biblical teaching has had considerable popularity in modern religious discussion of sexuality. The Anglican theologian Derrick Sherwin Bailey, for example, applies the term "henosis" to describe the "one flesh" spiritual union effected by sexual intercourse according to biblical thinking. Bailey finds an emphasis on such union in the Genesis account of the creation of male and female, and he traces its presence through later biblical texts ([1]; [3]). Although Bailey, like many thinkers, is uncomfortable with St. Paul's suggestion that even a chance liason with a prostitute can create such a one-flesh union ([1], pp. 51f.), he nevertheless finds this extreme remark by the Apostle generally consistent with biblical teaching in this area. Sexuality, according to this perspective, is filled with profound psychosomatic significance. Within the context of a committed, abiding personal relationship it can deepen and enhance human experience. Sundered from this context, it can lead to suffering (as partners are pulled unconsciously or against their will by the inner dynamics of sexual contact) or, if neglected, it can lead to a trivialization and debasement of sexual encounter and of our capacities for personal relationship ([16], Ch. 6).

It is no accident that this interpretation of biblical teaching has been popular in the modern period. Obviously, it enables biblically informed thinkers to validate sexuality in ways that go beyond the millennia-old insistence that sex cannot be separated from procreation. It also provides a basis for the traditional norms on sexuality that is less susceptible to changes in contraceptive technology than is the view based on procreative responsibility alone. But however useful this view may be to those who would defend biblical teachings, the question remains as to whether it is correct. Is it, in fact, true that sexual intercourse has the kinds of deep psychological implications attributed to it by those who defend this "henotic" view? And even if sexuality has significant relational dimensions, does it follow that the only appropriate context for sexual expression is in a lifelong personal relationship between the sexual partners?

Let me say frankly that I have no easy answers to these questions. It has struck me for some years that if biblical sexual theology has any enduring validity it is here. That human sexuality is not a matter of genital intercourse alone, that it can be charged with emotional and relational significance, seems to me true to many persons' experience (including my own). Nevertheless, even here there are good reasons for doubting the continuing validity or pertinence of this perspective. For one thing, it is not clear that sexual experience per se has the profound psychological implications this "henotic" perspective would accord it. I have already suggested, for example, that within the biblical tradition, the psychological entailments of sexuality are always connected with the possibility of conception and procreation and their social ramifications. It is interesting to note that the suggestion in Genesis that sexual intercourse renders a couple "one flesh" probably makes reference, not to any immediate psychological bond between the sexual pair, but to the offspring their union creates.

It is really true, therefore, that genital sexuality must have intimate relational meaning even when separated from any procreative significance? Indeed, in a context where sexuality really has been freed emotionally from its traditional procreative entailments, is sexual expression itself in any way the supremely relational experience that the "one flesh" view holds it to be? Or does it instead rank among a variety of possible human relational experiences without any particular pride of place? Is it even possible for sexuality to be incorporated into non-relational contexts without losing value and without truncating human emotional capacities? Once again, there are no easy answers to these questions. If the experience of many persons testifies on one side of the issue, the experience of others supports opposing views. I recall here, in particular, the statement of a woman interviewed in connection with a popular magazine's reportage on the life of "singles" in a major metropolitan area. This woman had become so accustomed to frequent sexual encounters – often on a first date – that sex had ceased to be an event of any special importance for her or even a measure of a special relationship with someone. Its place had been taken by the experience of sleeping with someone – literally sharing a bed for the night. She found herself willing to have this experience only with those individuals she especially valued. For her, this form of physical intimacy had become more a measure of caring than sexual intercourse.

While this kind of attitude might be held up as evidence of the debasement of sexuality through overuse of promiscuity, it can also be pointed to in support of the claim that sexuality is at best one – but not the only and not necessarily a consummate – area of interpersonal relationship. This suggests that the very strong view of sexuality's relational significance attributed to biblical theology – if it is in fact a biblical view – may not be altogether sustainable.

Beyond these considerations, there is also the question of whether any relational dimensions sexuality may have require it to be expressed in the context of the lifelong relationship of marriage. We are all familiar with the ways in which the prolongation of the human life span, changing social expectations and values, and the relative economic empowerment of women have worked together to vastly increase the divorce rate in modern societies. Critics may lament this phenomenon as a departure from the model of life-long relationship most suited to human well-being. But these phenomena may equally be read as a reasonable adaptation to changing realities. While "to death do us part" may have made sense in the context of traditional societies, where death's forced "parting" was often not far off for many young couples, it may make far less sense in a context where personal vigor can continue until the eighth decade of life and where persons are free to change and can be expected to change in very fundamental ways throughout the course of life. There is no doubt that many persons in our society value the context of commitment and support that matrimony represents. Why else is divorce so often followed by remarriage? But the fact that these same persons feel compelled to end marital relationships that have ceased to be rewarding suggests that the older ideal of lifelong commitment receives more lip-service than respect. If conduct is to be taken as a guide, it may be that today's widespread practice of "serial monogamy" through divorce and remarriage is no less suited to human psychological well-being than the older patterns.

Once again, I want to be careful. I am not saying that the perspectives attributed to biblical theology regarding marital commitment are wrong. My point is that they are not necessarily right. We are living in an era where, as far as sexuality is concerned, the earth has moved beneath us. Despite this, theologians and religious thinkers continue to reiterate, as uncontestable truths, teachings about human sexuality whose connections with a more traditional context are patent. It may be that some of these teachings remain valid despite the changes that have overtaken us. But it may also be that their application to a changed situation may create new forms of suffering as people try to mold their experience to an outdated normative model. In any case, those who would elaborate a theology of sexuality or a sexual ethic in our day must be open to the possibility that a venerated tradition is simply time-bound with respect to its insights in this important area.

(3) *That human sexuality is normatively heterosexual.* The point I am making about biblical teaching is even more graphically illustrated with regard to the issue of heterosexuality and homosexuality. There is no question that the Bible upholds, and rather sternly enforces, a view that the only permissible mode of human sexual expression is heterosexual. From the Genesis vision of creation through the Levitical prohibitions against homosexual conduct to the New Testament and St. Paul's castigations of homosexuals and homosexual behavior, biblical teaching

forms an unyielding charter for heterosexuality as the normative form of human sexual expression.

Once again it is not difficult to see why this is so. The social and cultural context of ancient Israel provided many factors that contribute to this viewpoint. Simple ignorance about homosexuals and homosexuality plays a role. The suggestion in Paul's writings that homosexuals are somehow free to "choose" their preferences and his apparent equation of all homosexual conduct with exploitative forms of pederasty are examples ([2], [22], [29]).

In the Hebrew Bible, a mistaken biology underlies a ban on any non-procreative "wastage" of semen. Also important is a possible connection between homosexual practices and the various fertility cults with which the religion of Israel was locked in mortal combat ([2], Ch. 2; [31]). Hebrew schemes of classification may be significant. As Mary Douglas has shown, Hebrew law is intensely concerned to order the world in terms of neat categories, and things that do not fit within or that bridge these categories, be they animals, foods, or sexual practices, are ordinarily condemned ([15], pp. 130, 139). The association of homosexuality with the practices of feared and hated military foes may also play a role ([22], p. 59). Finally, we cannot neglect the demographic realities of a small community on the edge of extinction, for which procreation is not just a blessing but a mandate. Any practice threatening the physical continuance of the community is naturally abhorred.

I should point out that a number of recent writers have tried to suggest that the biblical tradition – and the religious traditions to which it gives rise – may not be as unsparing on this issue as they are sometimes made out to be. It is frequently pointed out, for example, that the Sodom and Gomorrah episode, often taken as a proof-text for traditionalist interpretations, may deal with other matters, such as a failure of hospitality, rather than with homosexuality ([2], Ch. 1; [22], p. 59). Instances of male-male and female-female love (although not necessarily sexual love) in the Bible are also pointed to [20], and the castigations of homosexuality in the New Testament are related to the association of homosexual conduct with Rome's oppressive and coercive military institutions or to its association with pederasty [29]. Beyond this, it has also been pointed out that the Christian churches and the monastic communities often protected and provided a secure retreat from persecution for homosexual persons [9].

While this more recent scholarship is a healthy corrective to the view that biblical teaching is monolithic on this matter, nothing can change the fact that homosexuality is not regarded favorably within this tradition and that heterosexuality is viewed as the divinely constituted pattern for human sexual expression. Here, indeed, it can be said that popular "homophobic" attitudes (if not the persecution of homosexuals) find their cultural basis. Of course, for those who believe that biblical teaching on this and other matters constitutes revealed truth, the relatively unrelieved scriptural condemnation of homosexuality will point a way through this issue. Such persons will accept these condemnations on biblical grounds, and wherever possible, they may even seek to interpret the conflicting data of medicine and psychology in ways that support this biblical view (e.g., [32], [18]).

But many others will not agree. The outdated social and cultural bases of biblical teachings here are so obvious, their disconnection from any well founded scientific understanding of human sexual orientation (which we are only now beginning to develop) is so evident, that it is reasonable to put these teachings aside as we try today to understand these matters (and as we necessarily practice active respect for persons of different sexual orientation). The time has passed, I believe, when it is possible merely to "up-date" the tradition in this regard or to discover hidden resources and other meanings within it. A more honest approach is to acknowledge that it may flatly be wrong. Taking a cue from Nietzsche, it may be time to see the lightning and hear the thunder, which a generation of legitimate protest by a group of oppressed persons and a developing body of social scientific and sex research have helped create.

(4) *That human beings can be categorized and social roles can meaningfully be assigned on the basis of gender*. Feminist critics of the biblical tradition have sometimes characterized it as openly sexist and patriarchal and as deeply corrupted by these tendencies ([13], [14]). In contrast, those who adhere to a traditional "revelational" reading of biblical texts have sought to defend the forms of gender-related role ascription, hierarchy, and exclusion sanctioned by biblical teaching. Between these two factions stands a body of revisionist scholars who have tried to mitigate the presence of sexist elements in the Bible. Like those who have sought to reconsider biblical teaching on homosexuality, these scholars have tried to place some of the Bible's more offensive

gender-related teachings in context, and they have pointed to alternative views of this issue residing within the corpus of biblical teaching ([7], [25], [34]). Each of these approaches has some truth to it. Whether this is to be assessed as good or bad, there is no doubt that, in many ways, the Hebrew Bible and Christian New Testament furnish a charter for patriarchy. Thus, even as it locates it within the fallen order of sin, the Genesis cosmogony justifies an order of male dominion. The God of the Hebrew Bible, though without sex or gender, is typically regarded as having "male" attributes, and the prophets do not shrink from regarding his relationship to Israel as one of husband to wife, master to mistress (e.g., Hosea 2: 19; cf. Ephesians 5: 22ff.). Hebrew law perpetuates this sense of hierarchy through a normative structure that frequently subordinates women to men and that excludes them essentially from the vital religious center of community life. In the New Testament, largely in the writings of Paul and the pseudo-Pauline epistles, this structure of hierarchy and inferior role assignment is perpetuated. Indeed, in I Timothy: 14-15 reference is made to the Genesis text, and the order of creation, as a basis for women's submissive role. Nor is it insignificant that Christ himself is male and that the religiously most central female in the New Testament, Mary, finds her identity in motherhood.

On the other side, revisionist scholarship has called this reading of the Bible into question. Scholars have challenged the picture of female inferiority and domination, which generations of Jews and Christians have read into the Genesis text, and "female" qualities have been identified in the Hebrew conception of God ([28], pp. 99ff.; [34], Ch 2). The presence of independent, religiously significant women has been signalled in both the Old and New Testaments. And Paul's ringing assertion, in Galatians (3:28), that in the Christian community "there is neither Jew nor Greek, slave nor free, male nor female" has been taken as the charter for a new social vision that begins to find very partial embodiment in Christian monastic life in subsequent centuries.

While these revisionist efforts are impressive and a useful corrective to the revelational or critical, they often share with these views the assumption that Biblical teaching in this area is somehow religiously or morally significant. Not merely the defenders of a revelational approach to the Bible, but even some of its sharp critics tend to believe that these teachings are an integral aspect of biblical faith that must be taken seriously. Less to the fore is the view I am proposing: that on this matter, as on most others having to do with sexuality and gender, the Bible's teachings are simply irrelevant, that they are too outdated to merit significant attention. Certainly biblical writers could not be expected to have our experience of more than a century of efforts at female emancipation and equality, nor did they possess our growing understanding of the relative insignificance of gender as a marker of human abilities or capacities. While there may be glimmerings of such an understanding in the biblical materials, the fact remains that even revisionists here are more informed in their perspective by contemporary experience than they are by the Biblical texts themselves.

That the Bible is taken so seriously here by its critics and its friends is a reflection of the ongoing authority of its teachings in the areas of sexuality and gender. This in turn has a basis in the fact that the Bible is preoccupied with these matters, indeed, that it has almost overinvested its concern in them. But if we once acknowledge that this whole area of teaching, despite its salience, may be one of the least trustworthy and most outdated aspects of biblical theology and ethics, we may be able to gain greater distance from this aspect of the tradition. While allowing a role for sound historical and textual inquiry, this approach eliminates the need for apologetic efforts, preserves other areas of biblical teaching for inquiry regarding their continuing religious and moral validity, and frees us to regard our contemporary experience as the most important resource for a response to sex and gender-related issues.

I suspect that many persons not given to regarding each word of scripture as revealed would consider the fairly simple contention I have advanced as obvious. Yet it is remarkable how unwilling even sophisticated biblical theologians and ethicists have been to acknowledge that biblical teaching on sexuality may be thoroughly outdated. Instead, what has ruled theological discussion in this area has been a series of efforts to discover valid "core" teachings amidst the shards of outmoded biological or social conceptions. The shakiness of this approach is evidenced by the fact that such biblically-based sexual theologies have, in our time, proven to be remarkably perishable efforts. Either these theologies have served to perpetuate older errors in outwardly refurbished form and have come to appear ridiculous within a few years after their publication, or they have accommodated to contemporary information without fully acknowledging that they are essentially scrapping biblical teaching in the area at issue. Either way, and some discussions combine both of these tendencies, such efforts underscore the need to approach the tradition in a franker and more critical fashion.

Let me illustrate this point with brief reference to three important theological treatments of sexuality over the past generation: Karl Barth's discussion in his *Church Dogmatics* [4], Helmut Thielicke's *The Ethics of Sex* [33], and James B. Nelson's more recent book *Embodiment* [23]. These three books are representative of a much wider body of sexual-ethical and theological writing over the period they span ([1], [3], [8], [10], [11], [12], [16], [17], [19], [21]). Although these discussions are very different from one another, with Nelson's the least bound to the biblical text, they share the assumption that beneath a dross of culturally bound ideas and attitudes, something within this teaching remains important and normative for our thinking about sexuality today.

Barth's discussion of homosexuality in the *Church Dogmatics* clearly reveals the perils of this approach. As is true of many other matters that he tries to interpret in the light of biblical faith *and* contemporary culture, Barth's discussion is a sophisticated one. As a Protestant and a modern, he is wise enough to avoid interpreting various biblical condemnations of homosexual conduct in terms of older ideas of "natural law" and of the necessarily procreative intentionality of sex. Indeed, in place of this whole approach to sex, Barth puts forth a resoundingly "relational" and interpersonal view according to which sexuality is seen as a divinely appointed opportunity for human communion and self-transcendence. "Male and female being is the prototype of all I and Thou," he says, "of all the individuality in which man and man differ from and yet belong to each other" ([4], p. 150).

This view, though perhaps biblical, is deeply in touch with currents of modern attitudes on sexuality, and in view of this, one would expect some circumspection on Barth's part when it comes to the matter of treating homosexuality itself. But no such caution is evident. Speaking out of a context that precedes gay liberation, Barth simply assumes that his refurbishment of biblical teaching in one area need not carry over into another area where it has not yet been challenged. Indeed, Barth goes so far as to *apply* his updated sexual theology to the issue of homosexuality in a way that, with the hindsight of just three decades, must strike us as remarkably unconsidered and naive. Characterizing homosexuality as a "physical, a psychological and social sickness, the phenomenon of perversion, decadence and decay," he traces its origin to a refusal to admit the validity of the divine command to (heterosexual) fellow-humanity. Thus, homosexuality, for him, is an idolatrous form of self-worship, a rejection of the command to reach out to the other. "The real perversion takes place, the original decadence and disintegration begins," he affirms, "where man will not see his partner of the opposite sex and therefore the primal form of fellow-man, refusing to hear his question and to make a responsible answer, but trying to be human in himself as sovereign man or woman, rejoicing in himself in self-satisfaction and self-sufficiency" ([4], p. 16).

As many critics have pointed out, Barth's discussion here begs a number of key questions. Why, for example, is heterosexuality the "primal" form of relational fellow-humanity? If relationship and a reaching out beyond the self are a crucial implication of the divine command, why is not such reaching out between persons of the same sex as satisfactory as heterosexual relationality? And how can it be said that any reaching out to another human being is merely a form of self-satisfaction or self-adulation? Undoubtedly, clever interpreters of a Barthian view might wish to defend his claims further. But the fact that Barth does not even pause to consider the objections this view might arouse makes it clear that his position is less an insightful understanding of the nature and limits of human sexuality than it is a recasting of traditional religious opposition to homosexuality in updated theological terms. More than anything else, Barth reveals himself here to be a creature of his epoch.

In this respect, Barth's discussion illustrates not only the datedness of biblically-inspired teaching in this specific area, but also the dangers of trying to elaborate a theology of sexuality on the assumption that core teachings of the Bible remain valid while only peripheral ones require revision. The difficulty lies in determining what is central and what is peripheral. Certainly Barth himself was convinced – and rightly, I believe – that the Bible's condemnations of homosexuality are as central and as foundational as any such teachings can be. What this discussion illustrates, therefore, is that it may be far safer as we journey into uncharted areas of thinking about sexuality simply to assume that the entire tradition is untrustworthy in this area. Only by this strategy can we be genuinely open to new information and unbiased by views that trace their origin to now distant social and cultural circumstances.

With Helmut Thielicke's 1963 publication *The Ethics of Sex* we move ahead more than a decade in time, but similar problems are evidenced.

Indeed, while this volume is in some ways a step forward in Christian reflection about sexuality, in others it reveals attitudes even less advanced than Barth's. For example, where Barth justifies monogamy in terms of the divine command to abiding relationship in sexual expression, Thielicke justifies it in terms of alleged differences in male and female psychology and sexuality. It is the female's "sex nature", Thielicke tells us, that her personhood cannot be separated from the physical realm of sex, and as a result every sexual encounter marks her deeply and indelibly. This is not true of the male, however, and hence polygamy is a sexual possibility for the male but not for the female. Does this then justify polygynous marriage? No, Thielicke concludes, because Christian love and ethical consistency rule out the male's demanding for himself what he must deny to the woman ([33], pp. 80ff.).

I think it is reasonably clear that these alleged differences in male and female sexual nature owe their origin more to popular stereotypes shaped by prevalent cultural values than to any informed assessment of male or female sexuality. Once again we see the danger of proceeding from "core" biblical teachings – in this case the traditional Christian insistence on monogamy – in order to shape a contemporary perspective. When the biblical view is merely accepted as true, there develops a powerful tendency to marshall evidence to support this view, however doubtful that evidence's value.

This conclusion may seem less applicable to Thielicke's treatment of homosexuality, but it is pertinent there as well. To his credit, Thielicke questions Barth's assumption that homosexual relationships cannot be a Christian form of encounter with our fellow man ([33], p. 271). And though he defines homosexuality as fundamentally not in accord with the order of creation, and hence a sinful "perversion," he refuses to distinguish it from or raise it in significance above other forms of corruption that characterize human life since the Fall ([33], p. 283). In this context, homosexuality becomes an "ailment" like many others afflicting human beings, but an ailment which Thielicke concedes may be incurable and unchangeable. Hence, Christians are led to "accept" those who are homosexuals. Does this mean that an active homosexual lifestyle is Christianly permissible? Thielicke is not sure. On the one hand, in good Protestant fashion, he rejects celibacy for the homosexual no less than for the heterosexual because it is a "special calling" not open to most persons. But neither does he appear to wish to sanction full expression of homosexual tendencies. In a very tentative vein Thielicke asks whether the same norms might not apply here "as in the normal relationship of the sexes," and he concludes by suggesting that non-promiscuous, stable, and "monogamous" relationships between homosexual partners may be a permissible "exception" for those caught in an irreversible situation ([33], pp. 285-287).

Contrasted with Barth's, this is a relatively enlightened treatment of the issue. Indeed, if one puts aside the unjustifiable rhetoric about homosexuality as a "perversion" and violation of the order of creation, there are many who would agree with Thielicke's practical conclusion that "responsible" homosexual relationships may be morally and religiously acceptable. Yet to concede this is to miss the subtle way in which even here a traditional Christian and biblical agenda colors the discussion. For why must responsible homosexual expression follow the norm of "monogamous" heterosexual relationships? Why in view of possibly differing sexual needs of homosexuals from heterosexuals and apart from any possibility of procreation is it assumed that a traditional and heterosexual model of relationship must prevail? There may be answers to these questions, but Thielicke does not provide them, and he seems not even alert to the issue. While the circle of defensible "core" teachings has been narrowed, therefore, this circle remains and it continues in an almost unreflective way to shape Thielicke's sexual ethic.

Somewhat the same conclusion applies, along with others, to James Nelson's treatment of a variety of sexual issues in his book *Embodiment*. In fact, in many ways this is a remarkably contemporary discussion. Nelson is critical of the forms of sexism – or "sexist dualism" as he calls it – that mar the biblical tradition ([23], pp. 46, 58–69). He is open to the acceptability of forms of non-marital sexual expression, and of non-standard sexual variations. He is clear in his defense of homosexuals as persons and as Christians. Indeed, he is even willing to question the imposition of the "monogamous" model of a genitally exclusive, permanent relationship on homosexual conduct, asking whether this insistence really represents an attempt "to understand the range of meanings in homosexual relationships" ([23], p. 208).

Despite the contemporaneity of this discussion, however, it evidences two serious difficulties. First, even if this has been reduced to the narrowest circle, there remains a tenacious but still unquestioned adherence to elements of a biblical understanding of sexuality. "[P]articular biblical statements about human sexuality are inevitably historically conditioned," Nelson concedes. For this reason,

[S]pecific injunctions cannot legitimately be wrenched from their historical context and applied in a mechanical manner to the late twentieth century. Our essential scriptural guidance must come from the larger perspectives of biblical faith. It must come from the Bible's basic understanding of the human person in the light of God's presence, action and purpose ([23], pp. 51f.).

What, then, does this "basic understanding" imply where human sexuality is concerned? One answer seems to be that sexuality cannot be separated from a relational context. Criticizing views that miss the connection between erotic desire and intimate knowledge, he observes

The ancient Hebrews knew better when they occasionally used the verb "to know" (*yadah*) as a synonym for sexual intercourse. The sexual act at its best is the union of desiring and knowing. If I desire another sexually without wanting to have deep knowledge of the other, without wanting to be in a living communion with the partner, I am treating the other merely as object, as instrument, as means to my self-centered gratification. But in the union of desiring and knowing, the partner is treated as a self, the treasured participant in communion ([33], p. 33. Also pp. 86, 89).

Once again I have no wish to be placed in the position of saying that Nelson here, or the presumed biblical theology on which he draws, is mistaken. It may be true (as I personally believe) that sex "at its best" requires us to want a "deep knowledge of the other" and it may also be true (as I do not believe) that anything less in sexuality necessarily objectifies and injures the other. Nevertheless, is it not legitimate to ask whether any of this really is correct? In a discussion otherwise so up-to-date, so willing to question traditional insistences, why is no sustained effort made to evaluate a very contemporary view that sex can be exciting and fun, that even casual sex between consenting partners can have value and need not result in injury to the other (or at least not injury any more serious than other forms of casual relationship to persons)? Erica's Jong's "zipless fuck" may be a fantasy. But the fact that it is a fantasy for many persons opens up possibilities of sexual expression that merit consideration and attention. That one of the most progressive sexual theologians of our day cannot free himself here of the hold of traditional, biblically informed thinking testifies to the limits of this whole approach.

On another side, the limits of this approach reveal themselves in what, with apologies to an author I greatly respect, I can only call a subtle form of intellectual dishonesty. This involves once again the claim that conclusions with regard to sexual conduct that radically conflicts with numerous biblical teachings can somehow be justified in terms of *core* biblical teachings or what Nelson calls "the Bible's basic understanding of the human person in the light of God's presence, action and purpose." Among less acute writers, this understanding and approach very characteristically take form by making appeal to the elements of grace and forgiveness in the the Christian tradition. Typically, for example, writers straddle the explicit prohibitions on homosexual conduct found in the Bible, and their own sense that these prohibitions are not justified, by encouraging loving acceptance of the homosexual as a person or even toleration of homosexual conduct as one form of "fallen" human behavior ([4], pp. 166f). The dishonesty in this approach is highlighted by the frank fundamentalist objection that if something is a sin, it is a sin and should not be permitted. Surely no progressive theologian would approach genocide in this way by humanly accepting those who practice it and by classifying it among the uneliminable aspects of our fallen condition. Of course, homosexual conduct is not genocide and a more honest approach is to make clear one's explicit moral valuation of it.

Though he is not guilty of this kind of evasion, Nelson displays some subtle variations on it. For example, he prefaces his discussion of gayness by making reference to the central interpretive principle that "Jesus Christ is the bearer of God's invitation to human wholeness and communion" and hence "the central norm which and by which everything else in scripture should be judged" ([23], p. 181). But if this is true, if such a broad and amorphous norm as "human wholeness" is the chief criterion of right and wrong, why is it important to pay attention to any of the Bible's specific normative teachings on sexuality? Why not simply admit that we are on our own with regard to sexual ethics today?

Again, Nelson tells us that a further principle governing our thinking is that we should take seriously the historical context of the biblical writers and of ourselves. He relates this to the "radical monotheism" of Christian faith and to the related "Protestant Principle", which he interprets as an admonition not to absolutize any historically relative ethical and theological judgments ([23], pp. 133, 181). He also encourages Christians to consider the possibility that new insights on sexuality represent an aspect of God's ongoing revelation of His truth ([23], p. 181). Superficially regarded, Nelson here appears to be saying what I have tried to argue: that the Bible is merely outdated in its teaching in this area so that we must turn to contemporary sources for self-understanding and direction. But there is an important difference. The stance for which I have argued admits openly that this tradition may be wrong and it urges us to great caution in looking to the Bible for wisdom or instruction on this matter. Nelson's approach, however, is less frank and less cautious. The whole tenor of his remarks, for example, is that while certain Biblical teachings may be outdated, others, especially those affirming the value of human embodiment and of the relational dimensions of sexuality, are profoundly correct, are a genuine aspect of God's revelation. Thus, he continues to pick and choose among the resources of this tradition, and he continues to find clear religious support for normative directions he favors. The danger of this approach, as I have tried to suggest on the matter of relational sexuality, is that it perpetuates the tendency to sanctify norms and attitudes which a genuine openness to experience may not warrant.

I have no intention of saying that as we approach sexual questions in the future we must throw away the Bible, or the teachings of the traditions to which it has given rise. These do constitute sources of experience that merit our attention. My point, instead, is that the time has come to consider the possibility that despite its enormous investment in this area (indeed, perhaps because of its enormous investment in this area), the biblical tradition's teachings on sexuality, whether in whole or in part, may no longer be relevant. I will leave it to Jewish and Christian theologians to determine whether it is possible to elaborate a theology of sexuality that really takes this understanding into account and whether that theology can honestly claim to root itself in the Bible. Whatever the answers to these questions may be, the implications of what I have been saying for ethics seem clear: as we reconsider sexuality in our day, our first priority is to pay attention to the new experience our circumstances are creating, to approach this experience with honesty and without the fear that it may lead to conclusions that contradict long venerated assumptions and norms.

Dartmouth College,
Hanover, New Hampshire, U.S.A.

BIBLIOGRAPHY

1. Bailey, D. S.: 1952, *The Mystery of Love and Marriage*, Harper & Brothers, New York.
2. Bailey, D. S.: 1955, *Homosexuality and the Western Christian Tradition*, Longmans, Green, London.

3. Bailey, D. S.: 1959, *The Man-Woman Relation in Christian Thought*, Longmans, Green, London.
4. Barth, K.: 1951, 1961, *Church Dogmatics*, T & T Clark, Edinburgh.
5. de Beauvoir, S.: 1974, *The Second Sex*, H. M. Parshley (trans.), Vintage Books, New York.
6. Bertocci, P. A.: 1957, *Sex, Love and the Person*, Sheed and Ward, New York.
7. Bird, P.: 1974, 'Images of Women in the Old Testament', in R. Rosemary Ruether (ed.), *Religion and Sexism*, Simon and Schuster, New York, pp. 48–57.
8. Borowitz, E.: 1969, *Choosing a Sex Ethic: A Jewish Inquiry*, Schocken Books, New York.
9. Boswell, J.: 1980, *Christianity, Social Tolerance and Homosexuality*, University of Chicago Press, Chicago.
10. Brunner, E.: 1937, 1947, *The Divine Imperative*, Westminster Press, Philadephia.
11. Callahan, S.: 1968, *Beyond Birth Control: The Christian Experience of Sex*, Sheed and Ward, New York.
12. Curran, C. E.: 1978, *Issues in Sexual and Medical Ethics*, University of Notre Dame Press, Notre Dame.
13. Daly, M.: 1973, *Beyond God the Father*, Beacon Press, Boston.
14. Daly, M.: 1978, *Gyn/Ecology*, Beacon Press, Boston.
15. Douglas, M.: 1966, *Purity and Danger: An Analysis of Concepts of Pollution and Taboo*, Routledge & Kegan Paul, London.
16. Farley, M.: 1983, 'An Ethic for Same-Sex Relations', in R. Nugent (ed.), *A Challenge to Love*, Crossroad Publishing Company, New York, pp. 93–106.
17. Gordis, R.: 1978, *Love and Sex: A Modern Jewish Perspective*, Farrar Straus Giroux, New York.
18. Gordis, R.: 1980, 'Homosexuality and the Homosexual', in Edward Batchelor (ed.), *Homosexuality and Ethics*, Pilgrim Press, New York, pp. 52–60.
19. Haring, B.: 1966, *The Law of Christ*, Newman Press, Westminster.
20. Horner, T.: 1978, *Jonathan Loved David: Homosexuality in Biblical Times*, Westminster Press, Philadephia.
21. Kosnik, A., *et al.*: 1977, *Human Sexuality: New Directions in American Catholic Thought*, Paulist Press, New York.
22. McNeill, J. J., S. J.: 1976, *The Church and the Homosexual*, Sheed Andrews and McMeel, Kansas City.
23. Nelson, J.: 1978, *Embodiment*, Augsburg Publishing House, Minneapolis.
24. Nietzsche, F.: 1974, *The Gay Science*, Walter Kaufmann (trans.), Random House, New York.
25. Parvey, C. F.: 1974, 'The Theology and Leadership of Women in the New Testament', in R. Ruether (ed.), *Religion and Sexism*, Simon and Schuster, New York, pp. 117–149.
26. Patai, R.: 1960, *Family, Love, and the Bible*, Macgibbon and Kee, London.
27. Russell, B.: 1929, *Marriage and Morals*, Horace Liveright, London.
28. Russell, L. M.: 1973, *Human Liberation in a Feminist Perspective: A Theology*, Westminster Press, Philadephia.
29. Scroggs, R.: 1983, *The New Testament and Homosexuality*, Fortress Press, Philadelphia.
30. Smith, J. R. and Smith, L. G. (eds.): 1974, *Beyond Monogamy*, The Johns Hopkins Press, Baltimore.

31. Snaith, N. H.: 1967, *Leviticus and Numbers, The Century Bible*, Nelson, London.
32. Spero, M. H.: 1975, 'Homosexuality: Clinical and Ethical Challenges', *Tradition* **17**, 53–73.
33. Thielicke, H.: 1964, *The Ethics of Sex*, Harper & Row, New York.
34. Trible, P.: 1978, *God and the Rhetoric of Sexuality*, Fortress Press, Philadelphia.

NOTES ON CONTRIBUTORS

Robert Baker, Ph.D., is Associate Professor, Department of Philosophy, Union College Schenectady, New York.

Lisa Sowle Cahill, Ph.D., is Associate Professor, Department of Theology, Boston College, Chestnut Hill, Massachusetts.

Nancy N. Dubler, LL.B., is Director, Division of Legal and Ethical Issues in Health Care, Montefiore Medical Center, Bronx, New York.

John Duffy, Ph.D., is Professor Emeritus, Department of History, University of Maryland, College Park, Maryland.

H. Tristram Engelhardt, Jr., Ph.D., M.D., is Professor, Departments of Medicine and Community Medicine, and Member of the Center for Ethics, Medicine, and Public Issues, Baylor College of Medicine, Houston, Texas.

Mary Ann Gardell, M.A., is Managing Editor, *Journal of Medicine and Philosophy*, Center for Ethics, Medicine, and Public Issues, Baylor College of Medicine, Houston, Texas.

Joshua Golden, M.D., is Clinical Professor, and Director, Human Sexuality Program, Neuropsychiatric Hospital, Center for the Health Sciences, University of California, Los Angeles, California.

Ronald M. Green, Ph.D., is Professor, Department of Religion, Dartmouth College, Hanover, New Hampshire.

Nellie P. Grose, M.D., M.P.H., is Clinical Associate Professor of Family Medicine, Baylor College of Medicine, and in private practice, Houston, Texas.

Christie Hefner is President and Chief Operating Officer, Playboy Enterprises, Inc., Chicago, Illinois.

Sara Ann Ketchum, Ph.D., is Associate Professor of Philosophy, Department of Philosophy and Religion, Rollins College, Winter Park, Florida.

James J. McCartney, O.S.A., Ph.D., is Director, Bioethics Institute at St. Francis Hospital, Miami Beach, Florida.

J. Robert Meyners, Ph.D., is Director of Clinical Services, Masters and Johnson Institute, St. Louis, Missouri.

Earl E. Shelp, Ph.D., is Research Fellow, Institute of Religion, and Assistant Professor, Department of Community Medicine, and Member of the Center for Ethics, Medicine, and Public Issues, Baylor College of Medicine, Houston, Texas.

Paul D. Simmons, Ph.D., is Professor, Department of Christian Ethics, Southern Baptist Theological Seminary, Louisville, Kentucky.

Frederick Suppe, Ph.D., is Chairperson, Committee on the History and Philosophy of Science, University of Maryland, College Park, Maryland.

INDEX

The Philosophy and Medicine Book Series

Editors

H. Tristram Engelhardt, Jr. and Stuart F. Spicker

1. **Evaluation and Explanation in the Biomedical Sciences**
   1975, vi + 240 pp. ISBN 90–277–0553–4
2. **Philosophical Dimensions of the Neuro-Medical Sciences**
   1976, vi + 274 pp. ISBN 90–277–0672–7
3. **Philosophical Medical Ethics: Its Nature and Significance**
   1977, vi + 252 pp. ISBN 90–277–0772–3
4. **Mental Health: Philosophical Perspectives**
   1978, xxii + 302 pp. ISBN 90–277–0828–2
5. **Mental Illness: Law and Public Policy**
   1980, xvii + 254 pp. ISBN 90–277–1057–0
6. **Clinical Judgment: A Critical Appraisal**
   1979, xxvi + 278 pp. ISBN 90–277–0952–1
7. **Organism, Medicine, and Metaphysics**
   Essays in Honor of Hans Jonas on his 75th Birthday, May 10, 1978
   1978, xxvii + 330 pp. ISBN 90–277–0823–1
8. **Justice and Health Care**
   1981, xiv + 238 pp. ISBN 90–277–1207–7 (HB)/90–277–1251–4 (PB)
9. **The Law-Medicine Relation: A Philosophical Exploration**
   1981, xxx + 292 pp. ISBN 90–277–1217–4
10. **New Knowledge in the Biomedical Sciences**
    1982, xviii + 244 pp. ISBN 90–277–1319–7
11. **Beneficence and Health Care**
    1982, xvi + 264 pp. ISBN 90–277–1377–4
12. **Responsibility in Health Care**
    1982, xxiii + 285 pp. ISBN 90–277–1417–7
13. **Abortion and the Status of the Fetus**
    1983, xxxii + 349 pp. ISBN 90–277–1493–2
14. **The Clinical Encounter**
    1983, xvi + 309 pp. ISBN 90–277–1593–9
15. **Ethics and Mental Retardation**
    1984, xvi + 254 pp. ISBN 90–277–1630–7
16. **Health, Disease, and Causal Explanations in Medicine**
    1984, xxx + 250 pp. ISBN 90–277–1660–9
17. **Virtue and Medicine**
    Explorations in the Character of Medicine
    1985, xx + 363 pp. ISBN 90–277–1808–3
18. **Medical Ethics in Antiquity**
    Philosophical Perspectives on Abortion and Euthanasia
    1985, xxvi + 242 pp. ISBN 90–277–1825–3 (HB)/90–277–1915–2 (PB)
19. **Ethics and Critical Care Medicine**
    1985, xxii + 236 pp. ISBN 90–277–1820–2
20. **Theology and Bioethics**
    Exploring the Foundations and Frontiers
    1985, xxiv + 314 pp. ISBN 90–277–1857–1
21. **The Price of Health**
    1986, xxx + 280 pp. ISBN 90–277–2285–4
22. **Sexuality and Medicine**
    Volume I: Conceptual Roots
    1987, xxxii + 271 pp. ISBN 90–277–2290–0 (HB)/90–277–2386–9 (PB)
23. **Sexuality and Medicine**
    Volume II: Ethical Viewpoints in Transition
    1987, xxxii + 279 pp. ISBN 1–55608–013–1 (HB)/1–55608–016–6(PB)
24. **Euthanasia and the Newborn**
    Conflicts Regarding Saving Lives
    1987, xxx + 313 pp. ISBN 90–277–3299–4